Springer Undergraduate Mathematics Series

D1483494

Springer

London
Berlin
Heidelberg
New York
Barcelona
Hong Kong
Milan
Paris
Singapore
Tokyo

Advisory Board

P.J. Cameron *Queen Mary and Westfield College*
M.A.J. Chaplain *University of Dundee*
K. Erdmann *Oxford University*
L.C.G. Rogers *University of Bath*
E. Süli *Oxford University*
J.F. Toland *University of Bath*

Other books in this series

Analytic Methods for Partial Differential Equations *G. Evans, J. Blackledge, P. Yardley*
Applied Geometry for Computer Graphics and CAD *D. Marsh*
Basic Linear Algebra *T.S. Blyth and E.F. Robertson*
Basic Stochastic Processes *Z. Brzeźniak and T. Zastawniak*
Elements of Logic via Numbers and Sets *D.L. Johnson*
Elementary Number Theory *G.A. Jones and J.M. Jones*
Groups, Rings and Fields *D.A.R. Wallace*
Hyperbolic Geometry *J.W. Anderson*
Introduction to Laplace Transforms and Fourier Series *P.P.G. Dyke*
Introduction to Ring Theory *P.M. Cohn*
Introductory Mathematics: Algebra and Analysis *G. Smith*
Introductory Mathematics: Applications and Methods *G.S. Marshall*
Linear Functional Analysis *B.P. Rynne and M.A. Youngson*
Measure, Integral and Probability *M. Capiṅksi and E. Kopp*
Multivariate Calculus and Geometry *S. Dineen*
Numerical Methods for Partial Differential Equations *G. Evans, J. Blackledge, P.Yardley*
Sets, Logic and Categories *P. Cameron*
Topics in Group Theory *G. Smith and O. Tabachnikova*
Topologies and Uniformities *I.M. James*
Vector Calculus *P.C. Matthews*

GRACE LIBRARY CARLOW COLLEGE
PITTSBURGH PA 15213

Gareth A. Jones and J. Mary Jones

Information and Coding Theory

With 31 Figures

Q
360
J68
2000

Springer

CATALOGUED

Gareth A. Jones, MA, DPhil
Faculty of Mathematical Studies, University of Southampton,
Southampton SO17 1BJ, UK

J. Mary Jones, MA, DPhil
The Open University, Walton Hall, Milton Keynes MK7 6AA, UK

Cover illustration elements reproduced by kind permission of:
Aptech Systems, Inc., Publishers of the GAUSS Mathematical and Statistical System, 23804 S.E. Kent-Kangley Road, Maple Valley, WA 98038, USA. Tel: (206) 432 - 7855 Fax (206) 432 - 7832 email: info@aptech.com URL: www.aptech.com
American Statistical Association: Chance Vol 8 No 1, 1995 article by KS and KW Heiner 'Tree Rings of the Northern Shawangunks' page 32 fig 2
Springer-Verlag: Mathematica in Education and Research Vol 4 Issue 3 1995 article by Roman E Maeder, Beatrice Amrhein and Oliver Gloor 'Illustrated Mathematics: Visualization of Mathematical Objects' page 9 fig 11, originally published as a CD ROM 'Illustrated Mathematics' by TELOS: ISBN 0-387-14222-3, German edition by Birkhauser: ISBN 3-7643-5100-4.
Mathematica in Education and Research Vol 4 Issue 3 1995 article by Richard J Gaylord and Kazume Nishidate 'Traffic Engineering with Cellular Automata' page 35 fig 2. Mathematica in Education and Research Vol 5 Issue 2 1996 article by Michael Trott 'The Implicitization of a Trefoil Knot' page 14.
Mathematica in Education and Research Vol 5 Issue 2 1996 article by Lee de Cola 'Coins, Trees, Bars and Bells: Simulation of the Binomial Process' page 19 fig 3. Mathematica in Education and Research Vol 5 Issue 2 1996 article by Richard Gaylord and Kazume Nishidate 'Contagious Spreading' page 33 fig 1. Mathematica in Education and Research Vol 5 Issue 2 1996 article by Joe Buhler and Stan Wagon 'Secrets of the Madelung Constant' page 50 fig 1.

ISSN 1615-2085

ISBN 1-85233-622-6 Springer-Verlag London Berlin Heidelberg

British Library Cataloguing in Publication Data
Jones, Gareth A.
 Information and coding theory. - (Springer undergraduate
 mathematics series)
 1. Information theory 2. Coding theory
 I. Title II. Jones, J. Mary
 003.5'4
ISBN 1852336226

Library of Congress Cataloging-in-Publication Data
Jones, Gareth A.
 Information and coding theory / Gareth A. Jones and J. Mary Jones.
  p.   cm. -- (Springer undergraduate mathematics series)
 Includes bibliographical references and index.
 ISBN 1-85233-622-6 (alk. paper)
 1. Information theory. 2. Coding theory. I. Jones, J. Mary (Josephine Mary), 1946-
II. Title. III. Series.
Q360 .J68 2000
003'.54—dc21   00-030074

Apart from any fair dealing for the purposes of research or private study, or criticism or review, as permitted under the Copyright, Designs and Patents Act 1988, this publication may only be reproduced, stored or transmitted, in any form or by any means, with the prior permission in writing of the publishers, or in the case of reprographic reproduction in accordance with the terms of licences issued by the Copyright Licensing Agency. Enquiries concerning reproduction outside those terms should be sent to the publishers.

© Springer-Verlag London Limited 2000
Printed in Great Britain

The use of registered names, trademarks etc. in this publication does not imply, even in the absence of a specific statement, that such names are exempt from the relevant laws and regulations and therefore free for general use.

The publisher makes no representation, express or implied, with regard to the accuracy of the information contained in this book and cannot accept any legal responsibility or liability for any errors or omissions that may be made.

Typesetting: Camera ready by the author and Michael Mackey
Printed and bound at the Athenæum Press Ltd., Gateshead, Tyne & Wear
12/3830-543210 Printed on acid-free paper SPIN 10731522

Preface

As this Preface is being written, the twentieth century is coming to an end. Historians may perhaps come to refer to it as the century of information, just as its predecessor is associated with the process of industrialisation. Successive technological developments such as the telephone, radio, television, computers and the Internet have had profound effects on the way we live. We can see pictures of the surface of Mars or the early shape of the Universe. The contents of a whole shelf-load of library books can be compressed onto an almost weightless piece of plastic. Billions of people can watch the same football match, or can keep in instant touch with friends around the world without leaving home. In short, massive amounts of information can now be stored, transmitted and processed, with surprising speed, accuracy and economy.

Of course, these developments do not happen without some theoretical basis, and as is so often the case, much of this is provided by mathematics. Many of the first mathematical advances in this area were made in the mid-twentieth century by engineers, often relying on intuition and experience rather than a deep theoretical knowledge to lead them to their discoveries. Soon the mathematicians, delighted to see new applications for their subject, joined in and developed the engineers' practical examples into wide-ranging theories, complete with definitions, theorems and proofs. New branches of mathematics were created, and several older ones were invigorated by unexpected applications: who could have predicted that error-correcting codes might be based on algebraic curves over finite fields, or that cryptographic systems might depend on prime numbers?

Information Theory and Coding Theory are two related aspects of the problem of how to transmit information efficiently and accurately from a source, through a channel, to a receiver. This includes the problem of how to store information, where the receiver may be the same as the source (but later in time). As an example, space exploration has created a demand for accurate transmission of very weak signals through an extremely noisy channel: there is no point in sending a probe to Mars if one cannot hear and decode the messages it sends back. In its simplest form this theory uses elementary techniques from Probability Theory and Algebra, though later advances have been based on such topics as Combinatorics and Algebraic Geometry.

One important problem is how to compress information, in order to transmit it rapidly or store it economically. This can be done by reducing redundancy: a familiar example is the use of abbreviations and acronyms such as "UK", "IBM" and "radar" in place of full names, many of whose symbols are redundant from the point of view of information content. Similarly, we often shorten the names of our closest friends and relatives, so that William becomes Will or Bill.

Another important problem is how to detect and correct errors in information. Human beings and machines cannot be relied upon always to avoid mistakes, and if these are not corrected the consequences can be serious. Here the solution is to *increase* redundancy, by adding symbols which reinforce and protect the message. Thus the NATO alphabet Alpha, Bravo, Charlie, ..., used by armed forces, airlines and emergency services for spoken communication, replaces the letters A, B, C, ... with words which are chosen to sound as unlike each other as possible: for instance, B and V are often confused (they are essentially the same in some languages), but Victor is unlikely to be mistaken for Bravo, even when misheard as Bictor.

Information Theory, much of which stems from an important 1948 paper of Shannon [Sh48], uses probability distributions to quantify information (through the entropy function), and to relate it to the average word-lengths of encodings of that information. In particular, Shannon's Fundamental Theorem guarantees the existence of good error-correcting codes, and the aim of Coding Theory is to use mathematical techniques to construct them, and to provide effective algorithms with which to use them. Despite its name, Coding Theory does not involve the study of secret codes: this subject, Cryptography, is closely related to Information Theory through the concepts of entropy and redundancy, but since the mathematical techniques involved tend to be rather different, we have not included them.

This book, based on a third-year undergraduate course introduced at Southampton University in the early 1980s, is an attempt to explain the basic ideas of Information and Coding Theory. The main prerequisites are elementary Probability Theory and Linear Algebra, together with a little Calculus, as covered in a typical first-year university syllabus for mathematicians, engineers or scientists. Most textbooks in this area concentrate mainly or entirely on either Information Theory or Coding Theory. However, the two subjects are intimately related (through Shannon's Theorem), and we feel that there are strong arguments for learning them together, at least initially.

Chapters 1–5 (representing about 60% of the main text) are mainly on Information Theory. Chapter 1, which has very few prerequisites, shows how to
encode information in such a way that its subsequent decoding is unambiguous
and instantaneous: the main results here are the Sardinas–Patterson Theorem
(proved in Appendix A), and the Kraft and MacMillan inequalities, concerning
the existence of such codes. Chapter 2 introduces Huffman codes, which —
rather like Morse code — minimise average word-length by systematically assigning shorter code-words to more frequent symbols; here (as in Chapters 3–5)
we use some elementary Probability Theory, namely finite probability distributions. In Chapter 3 we use the entropy function, based on probabilities and
their logarithms, to measure information and to relate it, through a theorem
of Shannon, to the average word-lengths of encodings. Chapter 4 studies how
information is transmitted through a channel, possibly subject to distortion
by "noise" which may introduce errors; conditional probabilities allow us to
define certain system entropies, which measure information from several points
of view, such as those of the sender and the receiver. These lead to the concept
of channel capacity, which is the maximum amount of information a channel
can transmit. In Chapter 5 we meet Shannon's Fundamental Theorem, which
states that, despite noise, information can be transmitted through a channel
with arbitrarily great accuracy, at rates arbitrarily close to the channel capacity. We sketch a proof of this in the simple but important case of the binary
symmetric channel; a full proof for this channel, given in Appendix C, relies on
the only advanced result we need from Probability Theory, namely the Law of
Large Numbers, explained in Appendix B.

The basic idea of Shannon's Theorem is that one can transmit information
accurately by using code-words which are sufficiently unlike each other that,
even if some of their symbols are incorrect, the receiver is unlikely to confuse
them (think of Bravo and Victor). Unfortunately, neither the theorem nor its
proof shows us how to find specific examples of such codes, and this is the aim
of Coding Theory, the subject matter of Chapters 6 and 7. In these chapters,
which are rather longer than their predecessors, we introduce a number of fairly
simple examples of error-correcting codes. In Chapter 6 we use elementary, direct methods for this; the main result here is Hamming's sphere-packing bound,
which uses a simple geometric idea to give an upper bound on the number of
code-words which can correct a given number of errors. In Chapter 7 we construct slightly more advanced examples of error-correcting codes using Linear
Algebra and Matrix Theory, specifically the concepts of vector spaces and subspaces, bases and dimensions, matrix rank, and row and column operations.
We also briefly show how some ideas from Combinatorics and Geometry, such
as block designs and projective geometries, are related to codes.

The usual constraints of space and time have forced us to omit several interesting topics, such as the links with Cryptography mentioned above, and only briefly to sketch a few others. In Information Theory, for instance, Markov sources (those with a "memory" of previous events) appear only as an exercise, and similarly in Coding Theory we have not discussed cyclic codes and their connections with polynomial rings. Instead, we give some suggestions for further reading at the end of the book.

The lecture course on which this book is based follows Chapters 1–7, usually omitting Sections 5.5, 6.5, 6.6 and 7.5 and the Appendices. A course on Information Theory could be based on Chapters 1–5, perhaps with a little more material on Markov sources or on connections with Cryptography. A course on Coding Theory could follow Chapters 6 and 7, with some background material from Chapter 5 and some extra material on, for instance, cyclic codes or weight enumerators.

We have tried, wherever possible, to give credit to the originators of the main ideas presented in this book, and to acknowledge the sources from which we have obtained our results, examples and exercises. No doubt we have made omissions in this respect: if so, they are unintentional, and no slight was intended.

We are grateful to Keith Lloyd and Robert Syddall, who have both taught and improved the course on which this book is based, together with the hundreds of students whose reactions to the course have been so instructive. We thank Karen Barker, Beverley Ford, David Ireland and their colleagues at Springer for their advice, encouragement and expertise during the writing of this book. We are indebted to W.S. (further symbols are surely redundant) for providing the quotations which begin each chapter, and finally we thank Peter and Elizabeth for tolerating their occasionally distracted parents with unteenagerly patience and good humour.

Contents

Notes to the Reader

Chapters 1–5 cover the basic material on Information Theory, and they should be read in that order since each depends fairly heavily on its predecessors. The Sardinas–Patterson Theorem (§1.2) and Shannon's Fundamental Theorem (§5.4) are important results with rather long proofs; we have simply outlined the proofs in the text, and the complete proofs in Appendices A and C can be omitted on first reading since their details are not required later. Other sections not required later are §5.5, §6.5, §6.6 and §7.5.

In a sense, the book starts afresh with Chapters 6 and 7, which are about Coding Theory. These two chapters could be read on their own, though it would help to look first at some of Chapter 5, in particular §5.2 for the example of repetition codes, §5.3 for the concept of Hamming distance, and §5.4 and §5.6 for the motivation provided by Shannon's Theorem.

We assume familiarity with some of the basic concepts of Probability Theory (in Chapters 1–5) and Linear Algebra (in Chapters 6 and 7), together with a few results from Calculus; there is some suggested background reading on these topics at the end of the book, in the section *Suggestions for Further Reading*, together with some comments on further reading in Information and Coding Theory.

The exercises are an important feature of the book. Those embedded in the text are designed to test and reinforce the reader's understanding of the concepts immediately preceding them. Most of these are fairly straightforward, and it is best to attempt them right away, before reading further. The supplementary exercises at the end of each chapter are often more challenging; they may require several ideas from that chapter, and possibly also from earlier chapters. Some of these supplementary exercises are designed to encourage the reader to explore the theory further, beyond the topics we have covered. Solutions of all the exercises are given at the end of the book, but it is strongly recommended to try the exercises first before reading the solutions.

1
Source Coding

Words, words, words. (*Hamlet*)

This chapter considers how the information emanating from a source can be encoded, so that it can later be decoded unambiguously and without delay. These two requirements lead to the concepts of uniquely decodable and instantaneous codes. We shall find necessary and sufficient conditions for a code to have these properties, we shall see how to construct such codes, and we shall prove Kraft's and McMillan's inequalities, which essentially say that such codes exist if and only if they have sufficiently long code-words.

1.1 Definitions and Examples

Information theory is concerned with the transmission of information from a sender, through a channel, to a receiver. The sender and receiver could be people or machines. In most cases they are different, but when information is being stored for later retrieval, the receiver could be the sender at some future time. We will assume that the information comes from a *source* S, which emits a sequence $\mathbf{s} = X_1 X_2 X_3 \ldots$ of symbols X_n; for instance, X_n might be the n-th symbol in some message, or the outcome of the n-th repetition of some experiment. In practice, this sequence will always be finite (nothing lasts for ever), but for theoretical purposes it is sometimes useful also to consider infinite sequences. We will assume that each symbol X_n is a member of some fixed finite set $S = \{s_1, \ldots, s_q\}$, called the *source alphabet* of S. For simplicity we will also assume that the probability $\Pr(X_n = s_i)$, that the n-th symbol X_n is s_i, depends only on i but not on n, so we write

$$\Pr(X_n = s_i) = p_i$$

for $i = 1, \ldots, q$. Thus different symbols may have different probabilities, but these remain constant in time (so S is *stationary*), and do not depend on the preceding symbols X_m where $m < n$ (so S is *memoryless*). In more advanced forms of this theory, such factors are taken into consideration, but we will ignore them here. As with any probability distribution, the probabilities p_i must satisfy

$$p_i \geq 0 \quad \text{and} \quad \sum_{i=1}^{q} p_i = 1 \tag{1.1}$$

(so each $p_i \leq 1$). In statistical terms, one can regard S as a sequence of independent, identically distributed random variables X_n, with probability distribution (p_i).

Example 1.1

S is an unbiased die[1], $S = \{1, \ldots, 6\}$ with $q = 6$, $s_i = i$ for $i = 1, \ldots, 6$, X_n is the outcome of the n-th throw, and $p_i = 1/6$ for $i = 1, \ldots, 6$. A biased die is similar, but with different probabilities p_i.

Example 1.2

S is the weather at a particular place, with X_n representing the weather on day n. For simplicity, we could let S consist of $q = 3$ types of weather (good, moderate and bad, for instance), so p_i $(i = 1, 2, 3)$ is the probability of each type, say $p_1 = 1/4$, $p_2 = 1/2$, $p_3 = 1/4$. (Here we ignore seasonal variations, which may cause the probability distribution (p_i) to vary in time.)

Example 1.3

S is a book, S consists of all the symbols used (letters, punctuation marks, numerals, etc.), X_n is the n-th symbol in the book, and p_i is the frequency of the i-th symbol in the source alphabet. (Here we ignore the effect of preceding symbols on probabilities: for instance, in English text the symbol "q" is almost always followed by "u".)

To encode a source, we use a finite *code alphabet* $T = \{t_1, \ldots, t_r\}$ consisting of r *code-symbols* t_j. In general, this is distinct from the source alphabet $S =$

---

[1] The singular of dice, as in "the die is cast".

$\{s_1, \ldots, s_q\}$, since it depends on the technology of the channel rather than the source. We call r the *radix* (meaning "root"), and we refer to the code as an *r-ary* code. In many examples, $r = 2$ and the code is called *binary*. Most binary codes, such as ASCII (used in computing), have $T = \mathbf{Z}_2 = \{0, 1\}$, the set of integers mod (2). Codes with $r = 3$ are called *ternary*. We encode S by assigning a code-word w_i (a finite sequence of code-symbols) to each symbol $s_i \in S$; to encode $\mathbf{s} = X_1 X_2 X_3 \ldots$ we represent each $X_n = s_i$ by its code-word w_i, giving a sequence $\mathbf{t}$ of symbols from T. For conciseness, we do not separate the code-words in $\mathbf{t}$ with punctuation marks or blanks; if these are used, they must be regarded as elements of T appearing at the beginning or end of each w_i. Thus Morse, which appears to be binary, is actually a ternary code: the three symbols are $\bullet$, — and a blank.

Example 1.4

If S is an unbiased die, as in Example 1.1, take $T = \mathbf{Z}_2$ and let w_i be the binary representation of the source-symbol $s_i = i$ $(i = 1, \ldots, 6)$. Thus $w_1 = 1$, $w_2 = 10, \ldots, w_6 = 110$, so a sequence of throws such as $\mathbf{s} = 53214$ is encoded as $\mathbf{t} = 10111101100$.

For clearer exposition, we will occasionally break our rule not to use punctuation, and insert full stops or brackets to show how $\mathbf{t}$ is decomposed into code-words: in Example 1.4 we could write $\mathbf{t} = 101.11.10.1.100$, for instance. This is purely for the reader's benefit, and the punctuation symbols should not be regarded as part of $\mathbf{t}$.

We need to define codes more precisely. A *word* w in T is a finite sequence of symbols of T, its *length* $|w|$ is the number of symbols. The set of all words in T is denoted by T^*; this includes the *empty word*, of length 0, which we will denote by ε. The set of all non-empty words in T is denoted by T^+. Thus

$$T^* = \bigcup_{n \geq 0} T^n \quad \text{and} \quad T^+ = \bigcup_{n > 0} T^n,$$

where $T^n = T \times \cdots \times T$ (with n factors) is the set of words of length n. A *source code* (or simply a *code*) $\mathcal{C}$ is a function $S \to T^+$, that is, an assignment of code-words $w_i = \mathcal{C}(s_i) \in T^+$ to the symbols $s_i \in S$. Many properties of codes depend only on the code-words w_i, and not on the particular correspondence between them and the symbols s_i, so we will often regard $\mathcal{C}$ as simply a finite set of words $w_1, \ldots, w_q$ in T^+. If S^* is defined by analogy with T^*, then one can extend $\mathcal{C}$ to a function $S^* \to T^*$ in the obvious way, encoding each $\mathbf{s}$ in S^* by using $\mathcal{C}$ to encode its successive symbols:

$$\mathbf{s} = s_{i_1} s_{i_2} \ldots s_{i_n} \mapsto \mathbf{t} = w_{i_1} w_{i_2} \ldots w_{i_n} \in T^*.$$

The image of this function is the set

$$C^* = \{w_{i_1} w_{i_2} \ldots w_{i_n} \in T^* \mid \text{each } w_{i_j} \in \mathcal{C}, \ n \geq 0\}.$$

We denote the length $|w_i|$ of w_i by l_i, so each $l_i \geq 1$. The *average word-length* of $\mathcal{C}$ is

$$L(\mathcal{C}) = \sum_{i=1}^{q} p_i l_i. \tag{1.2}$$

Example 1.5

The code $\mathcal{C}$ in Example 1.4 has $l_1 = 1$, $l_2 = l_3 = 2$ and $l_4 = l_5 = l_6 = 3$, so

$$L(\mathcal{C}) = \frac{1}{6}(1 + 2 + 2 + 3 + 3 + 3) = \frac{7}{3}.$$

The aim is to construct codes $\mathcal{C}$ for which

(a) there is easy and unambiguous decoding $\mathbf{t} \mapsto \mathbf{s}$,

(b) the average word-length $L(\mathcal{C})$ is small.

The rest of this chapter considers criterion (a), and the next chapter considers (b).

1.2 Uniquely Decodable Codes

A code $\mathcal{C}$ is *uniquely decodable* (= *uniquely decipherable*, or *u.d.* for short) if each $\mathbf{t} \in T^*$ corresponds under $\mathcal{C}$ to at most one $\mathbf{s} \in S^*$; in other words, the function $\mathcal{C} : S^* \to T^*$ is one-to-one, so each $\mathbf{t}$ in its image C^* can be decoded uniquely. We will always assume that the code-words w_i in $\mathcal{C}$ are distinct, for if $w_i = w_j$ with $i \neq j$ then $\mathbf{t} = w_i$ could represent either s_i or s_j, so $\mathcal{C}$ is not uniquely decodable. Under this assumption, the definition of unique decodability of $\mathcal{C}$ is that whenever

$$u_1 \ldots u_m = v_1 \ldots v_n$$

with $u_1, \ldots, u_m, v_1, \ldots, v_n \in \mathcal{C}$, we have $m = n$ and $u_i = v_i$ for each i. In algebraic terms we are saying that each code-sequence $\mathbf{t} \in C^*$ can be factorised in a unique way as a product of code-words.

Example 1.6

In Example 1.4, the binary coding of a die is not uniquely decodable: for instance, $\mathbf{t} = 11$ could be decomposed into code-words as 1.1 or 11, representing $\mathbf{s} = 11$ or $\mathbf{s} = 3$ (two throws of 1, or one throw of 3). We could remedy this by using a different code, with 3-digit binary representations of the source-symbols:

$$1 \mapsto 001, \ 2 \mapsto 010, \ \ldots \ , \ 6 \mapsto 110.$$

Then $\mathbf{s} = 11 \mapsto \mathbf{t} = 001001$ whereas $\mathbf{s} = 3 \mapsto \mathbf{t} = 011$. More generally, we have:

Theorem 1.7

If the code-words w_i in $\mathcal{C}$ all have the same length, then $\mathcal{C}$ is uniquely decodable.

Proof

Let l be the common word-length. If some $\mathbf{t} \in C^*$ factorises as $u_1 \ldots u_m = v_1 \ldots v_n$ with each $u_i, v_j \in C$, then $lm = |\mathbf{t}| = ln$, so $m = n$. Now u_1 and v_1 both consist of the first l symbols of $\mathbf{t}$, so $u_1 = v_1$, and similarly $u_i = v_i$ for all i. $\square$

If all the code-words in $\mathcal{C}$ have the same length l, we call $\mathcal{C}$ a *block code* of length l. We will study such codes in detail in Chapters 5–7. The converse of Theorem 1.7 is false:

Example 1.8

The binary code $\mathcal{C}$ given by

$$s_1 \mapsto w_1 = 0, \ s_2 \mapsto w_2 = 01, \ s_3 \mapsto w_3 = 011$$

has variable lengths, but is still uniquely decodable. Within $\mathbf{t}$, each symbol 0 indicates the start of a code-word w_i, and i is 1 plus the number of subsequent 1s. For instance,

$$\mathbf{t} = 001011010011 = 0.01.011.01.0.011 \quad \Rightarrow \quad \mathbf{s} = s_1 s_2 s_3 s_2 s_1 s_3.$$

In effect, we are using the symbol $0 \in T$ here as a punctuation mark.

We are going to state a necessary and sufficient condition for a code $\mathcal{C}$ to be uniquely decodable. We use induction to define a sequence $\mathcal{C}_0, \mathcal{C}_1, \ldots$ of sets of non-empty words, so $\mathcal{C}_n \subseteq T^+$ for all n. Specifically, we define $\mathcal{C}_0 = \mathcal{C}$, and

$$\mathcal{C}_n = \{ w \in T^+ \mid uw = v \text{ where } u \in \mathcal{C}, \, v \in \mathcal{C}_{n-1} \text{ or } u \in \mathcal{C}_{n-1}, \, v \in \mathcal{C} \} \quad (1.3)$$

for each $n \geq 1$; we then define

$$C_\infty = \bigcup_{n=1}^{\infty} C_n. \tag{1.4}$$

This definition may look a little daunting at first, but it should become clearer if we take it step by step: we start with $C_0 = C$, we then construct each C_n $(n \geq 1)$ in terms of its predecessor C_{n-1}, and finally we take $C_\infty = C_1 \cup C_2 \cup \cdots$. Note that for $n = 1$ the definition of C_n can be simplified: since $C_{n-1} = C_0 = C$ the two conditions separated by the word "or" in (1.3) are identical, so

$$C_1 = \{ w \in T^+ \mid uw = v \text{ where } u, v \in C \}.$$

Note also that if $C_{n-1} = \emptyset$ then $C_n = \emptyset$, so iterating this gives $C_{n+1} = C_{n+2} = \cdots = \emptyset$.

Example 1.9

Let $C = \{0, 01, 011\}$ as in Example 1.9. Then (1.3) gives $C_1 = \{1, 11\}$: we have $1 \in C_1$ since $01.1 = 011$ with $01, 011 \in C = C_0$, and $11 \in C_1$ since $0.11 = 011$ with $0, 011 \in C = C_0$. At the next stage, with $n = 2$, inspection shows that there is no $w \in T^+$ satisfying $uw = v$ where $u \in C$, $v \in C_1$ or vice versa. Thus $C_2 = \emptyset$, so $C_n = \emptyset$ for all $n \geq 2$ and hence $C_\infty = C_1 = \{1, 11\}$ by (1.4).

From the definition of C_∞, it is conceivable that the construction of this set might take infinitely many steps, requiring a new set C_n to be constructed for each $n \geq 1$. Exercise 1.1 shows that one can always construct C_∞ in finitely many steps, as in Example 1.9.

Exercise 1.1

Prove that if C has code-words of lengths $l_1, \ldots, l_q$, and $w \in C_n$ for some n, then $|w| \leq l = \max(l_1, \ldots, l_q)$. Deduce that each C_n is finite, and the sequence of sets $C_0, C_1, \ldots$ is eventually periodic. How does this help in the construction of C_∞ ?

Exercise 1.2

Construct the sets C_n and C_∞ for the ternary code $C = \{02, 12, 120, 20, 21\}$. Do the same for $C = \{02, 12, 120, 21\}$.

We can now give a necessary and sufficient condition for unique decodability. The *Sardinas–Patterson Theorem* [SP53] is as follows.

Theorem 1.10

A code C is uniquely decodable if and only if the sets C and C_∞ are disjoint.

Before considering a proof, let us apply this result in some simple cases.

Example 1.11

If $C = \{0, 01, 011\}$ as in Examples 1.8 and 1.9, then $C_\infty = \{1, 11\}$ which is disjoint from C. It follows from Theorem 1.10 that C is uniquely decodable, as we have already seen.

Example 1.12

Let C be the ternary code $\{01, 1, 2, 210\}$. Using (1.3) we find that $C_1 = \{10\}$, $C_2 = \{0\}$ and $C_3 = \{1\}$, so $1 \in C \cap C_\infty$ and thus C is not uniquely decodable (there is no need to calculate C_n for $n > 3$). To find an example of non-unique decodability, note that $10 \in C_1$ since $2 \in C$ and $2.10 = 210 \in C$, then $0 \in C_2$ since $1 \in C$ and $1.0 = 10 \in C_1$, and then $1 \in C_3$ since $0 \in C_2$ and $0.1 = 01 \in C$. Putting these equations together we get

$$210.1 = 2.10.1 = 2.1.0.1 = 2.1.01,$$

so the code-sequence $\mathbf{t} = 2101$ can be decoded as 210.1 or as $2.1.01$.

Exercise 1.3

Determine whether or not the codes $C = \{02, 12, 120, 20, 21\}$ and $C = \{02, 12, 120, 21\}$ considered in Exercise 1.2 are uniquely decodable. If C is not uniquely decodable, find a code-sequence which can be decoded in at least two ways.

Since the proof of the Sardinas–Patterson Theorem is rather long, we will give it in Appendix A; here instead, we will give two typical arguments to illustrate the ideas involved.

($\Rightarrow$) Suppose that $C \cap C_\infty \neq \emptyset$, say $w \in C \cap C_2$; thus $uw = v$ with $u \in C$ and $v \in C_1$ or vice versa, and for simplicity let us assume that the first case holds (see Exercise 1.4 for the second case). Then $u'v = v'$ where $u', v' \in C$, so the sequence $\mathbf{t} = u'uw \in T^*$ could represent a sequence $\mathbf{s}$ of three source-symbols (since $u', u, w \in C$) or one source-symbol (since $u'uw = u'v = v' \in C$). Thus decoding is not unique.

($\Leftarrow$) Suppose that we have an instance of non-unique decoding of the form $\mathbf{t} = u_1 u_2 = v_1 v_2$, where $u_1, u_2, v_1, v_2 \in C$. We cannot have $|u_1| = |v_1|$, for this would give $u_1 = v_1$ and so $u_2 = v_2$. Renumbering if necessary, we may therefore assume that $|u_1| > |v_1|$, so $u_1 = v_1 w$ where $|w| > 0$. Then $w \in C_1$, so $u_2 \in C_2$ since $wu_2 = v_2$. Thus $u_2 \in C \cap C_\infty$, so C and C_∞ are not disjoint.

Exercise 1.4

Suppose that $w \in C \cap C_2$, where $uw = v$ with $u \in C_1$ and $v \in C$. Give an example of a code-sequence which can be decoded in more than one way.

Exercise 1.5

A code C exhibits non-unique decodability in the form $012120.120 = 01.212.01.20$; find an element of $C \cap C_\infty$.

Exercise 1.6

Suppose that $w \in C \cap C_3$. By considering the various reasons why one could have $w \in C \cap C_3$, give examples of code-sequences which cannot be decoded uniquely.

The general arguments in the proof of Theorem 1.10 are similar to those outlined above, but they are rather more complicated since they have to deal with infinitely many different cases. Fortunately, there is a simpler necessary and sufficient condition for another important type of code, which we will consider in the next section.

We have defined unique decodability to mean that all finite code-sequences t can be decoded uniquely, but one could also consider the stronger requirement that this should be true for *all* code-sequences, finite or infinite. A theorem due to Even [Ev63], Levenshtein [Le64] and Riley [Ri67] shows that this happens if and only if $C \cap C_\infty = \emptyset$ and $C_n = \emptyset$ for some $n \geq 1$. (These are also necessary and sufficient conditions for C to be *uniquely decodable with bounded delay*, meaning that there is a constant d such that if two code-sequences agree in their first d symbols, then they have the same first code-word; thus decoding can begin after a delay of at most d symbols. We will consider a stronger condition in the next section.)

Example 1.13

In Example 1.9 above, both conditions are satisfied, so all code-sequences are decoded uniquely. For an example where all finite code-sequences are decoded uniquely, but some infinite ones are not, see Exercise 1.7.

Exercise 1.7

For each of the ternary codes C in Exercise 1.2, determine whether or not all infinite code-sequences can be decoded uniquely. If not, give an example of such non-unique decoding.

For the remainder of this book, we will restrict our attention to finite code-sequences.

1.3 Instantaneous Codes

Before defining instantaneous codes, let us consider a few examples.

Example 1.14

Consider the binary code C given by

$$s_1 \mapsto 0, \; s_2 \mapsto 01, \; s_3 \mapsto 11.$$

Using the notation of §1.2, we have $C_1 = C_2 = \cdots = \{1\}$, so $C_\infty = \{1\}$; thus $C \cap C_\infty = \emptyset$, so C is uniquely decodable by Theorem 1.10. Now suppose that we receive a finite message beginning $\mathbf{t} = 0111\ldots$. Although we know that this can be decoded uniquely, we cannot start to decode it until we come to the end of the block of consecutive 1s: if the number of 1s in this block is even, the decomposition of $\mathbf{t}$ must be $0.11.11.11\ldots$, and the decoded message must begin $\mathbf{s} = s_1 s_3 s_3 s_3 \ldots$; however, if the number of 1s is odd, the decomposition must be $01.11.11.11\ldots$, so $\mathbf{s} = s_2 s_3 s_3 s_3 \ldots$. In a practical situation, this delay in decoding could cause difficulties. We say that C is not instantaneous.

Example 1.15

The Prime Minister's telex prints RUSSIANS DECLARE WAR ... ; a quick decision is made, a button is pressed, and within minutes there are some very loud explosions. Soon, everyone is dead. Meanwhile, the telex continues printing ... RINGTON VODKA TO BE EXCELLENT.

Exercise 1.8

Show that the binary code $C = \{0, 01, 011, 111\}$ is uniquely decodable; how should the receiver react on receiving a sequence starting $0111\ldots1\ldots$?

Example 1.16

Consider the binary code $\mathcal{D}$ given by

$$s_1 \mapsto 0, \; s_2 \mapsto 10, \; s_3 \mapsto 11,$$

the reverse of the code C in Example 1.14. We can see this is uniquely decodable, either by Theorem 1.10, or because C is. It is also instantaneous, in the sense that we can decode a received message $\mathbf{t}$ as we go along: a 0 indicates w_1, which we decode as s_1, and a 1 indicates the start of $w_2 = 10$ or $w_3 = 11$, decoded as s_2 or s_3 as soon as we know the next symbol. Thus any code-word in $\mathbf{t}$ can be decoded as soon as it arrives, without delay.

Exercise 1.9

Is this also true for the code $\mathcal{D} = \{0, 10, 110, 111\}$, the reverse of the code $\mathcal{C}$ in Exercise 1.8?

Now for the formal definition: a code $\mathcal{C}$ is *instantaneous* if, for each sequence of code-words $w_{i_1}, w_{i_2}, \ldots, w_{i_n}$, every code-sequence beginning $\mathbf{t} = w_{i_1} w_{i_2} \ldots w_{i_n} \ldots$ is decoded uniquely as $\mathbf{s} = s_{i_1} s_{i_2} \ldots s_{i_n} \ldots$, no matter what the subsequent symbols in $\mathbf{t}$ are. Thus the code $\mathcal{C}$ in Example 1.14 is not instantaneous: a sequence $\mathbf{t} = w_1 w_3 \ldots = 011 \ldots$ might be decoded as $\mathbf{s} = s_1 s_3 \ldots$ or as $s_2 s_3 \ldots$, depending on the subsequent symbols. The code $\mathcal{D}$ in Example 1.16 is instantaneous: once $w_{i_1} w_{i_2} \ldots w_{i_n}$ is received, we know that it represents $s_{i_1} s_{i_2} \ldots s_{i_n}$, regardless of what comes next. By definition, every instantaneous code is uniquely decodable; Example 1.14 shows that the converse is false.

A code $\mathcal{C}$ is a *prefix code* if no code-word w_i is a prefix (initial segment) of any code-word w_j $(i \neq j)$; equivalently, $w_j \neq w_i w$ for any $w \in T^*$, that is, $\mathcal{C}_1 = \emptyset$ in the notation of §1.2. Thus $\mathcal{C}$ is not a prefix code in Example 1.14 (since 0 is a prefix of 01), but the reversed code $\mathcal{D}$ in Example 1.16 is a prefix code.

Theorem 1.17

A code $\mathcal{C}$ is instantaneous if and only if it is a prefix code.

Proof

($\Rightarrow$) If $\mathcal{C}$ is not a prefix code, say w_i is a prefix of w_j, then a code-sequence beginning $\mathbf{t} = w_i \ldots$ might be decoded as $\mathbf{s} = s_i \ldots$ or as $\mathbf{s} = s_j \ldots$, so $\mathcal{C}$ is not instantaneous.

($\Leftarrow$) If $\mathcal{C}$ is a prefix code, and $\mathbf{t}$ starts with $w_i \ldots$, then $\mathbf{s}$ must start with s_i, since no code-word w_j $(j \neq i)$ is a prefix of w_i or has w_i as a prefix. We can continue like this, decoding successive code-words in $\mathbf{t}$ as we receive them, so $\mathcal{C}$ is instantaneous.   □

Examples 1.14 and 1.16 are illustrations of this result.

1.4 Constructing Instantaneous Codes

In order to understand the construction of instantaneous codes, it is useful to regard the set T^* of words in T as a graph, that is, a set of points (called *vertices*), some pairs of which are joined by edges. In this case, the vertices are the words $w \in T^*$, and each w is joined by an edge to the r words $wt_1, \ldots, wt_r$ formed by adding a single symbol $t_i \in T$ to the end of w. One can visualise this graph as growing upwards, with the empty word ε at the bottom, and the words of length l at level l above ε; in Graph Theory, such a graph is called an *r-ary rooted tree*. (A *tree* is a connected graph with no circuits; here the root is ε.) Figure 1.1 shows the binary tree T^*, up to level $l = 3$, with $T = \mathbf{Z}_2$.

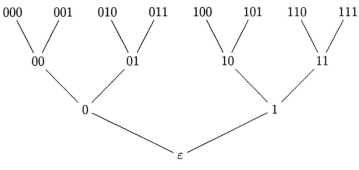

Figure 1.1

Exercise 1.10

Draw the ternary tree T^*, up to level $l = 3$, with $T = \mathbf{Z}_3$.

A code $\mathcal{C}$ can be regarded as a finite set of vertices of the tree T^*. A word w_i is a prefix of w_j if and only if the vertex w_i is dominated by the vertex w_j, that is, there is an upward path in T^* from w_i to w_j, so it follows from Theorem 1.17 that $\mathcal{C}$ is instantaneous if and only if no vertex $w_i \in \mathcal{C}$ is dominated by a vertex $w_j \in \mathcal{C}$ $(i \neq j)$. We can use this criterion to construct instantaneous codes, choosing vertices in T^* one at a time so that no vertex dominates (or is dominated by) any predecessor.

Example 1.18

Let us find an instantaneous binary code $\mathcal{C}$ for a source $\mathcal{S}$ with five symbols $s_1, \ldots, s_5$. First let us try $s_1 \mapsto w_1 = 0$, so 0 is a vertex in $\mathcal{C}$. If $\mathcal{C}$ is to be a prefix code, then no other vertex in $\mathcal{C}$ can dominate 0, so they must all dominate 1 (i.e. the other code-words must begin with 1). If we try $s_2 \mapsto w_2 = 1$ then no further code-words can be added, since they would dominate w_1 or w_2. Instead, let us try $s_2 \mapsto w_2 = 10$. Then $s_3 \mapsto w_3 = 11$ is impossible, since it allows no further code-words, so let us try $s_3 \mapsto w_3 = 110$. Continuing, we find the possibility $s_4 \mapsto w_4 = 1110$, $s_5 \mapsto w_5 = 1111$. This gives an instantaneous binary code $\mathcal{C} = \{0, 10, 110, 1110, 1111\}$ with word-lengths $l_i = 1, 2, 3, 4, 4$, shown in Figure 1.2. (This is not the only possibility: for instance, the binary code $\{00, 01, 10, 110, 111\}$ is also instantaneous.)

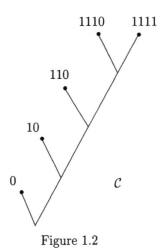

Figure 1.2

Example 1.19

Is there an instantaneous binary code for this source S with word-lengths $1, 2, 3, 3, 4$? Again, we use the binary tree T^*. Any choice of a code-word w_1 of length $l_1 = 1$, that is, a vertex of height 1, eliminates half of the vertices in T^* as possible code-words $w_2, \ldots, w_5$, namely all those dominating w_1, so a proportion $1 - \frac{1}{2} = \frac{1}{2}$ remains. A choice of w_2 at height 2 eliminates a further $1/2^2 = 1/4$ of T^*, leaving a proportion $1 - \frac{1}{2} - \frac{1}{4} = \frac{1}{4}$. Any choices of w_3 and w_4 at level 3 eliminate $1/2^3 + 1/2^3 = 1/4$ of T^*, so no vertices are left for w_5. The difficulty is that the sum of the proportions of T^* above each w_i exceeds 1:

$$\frac{1}{2} + \frac{1}{2^2} + \frac{1}{2^3} + \frac{1}{2^3} + \frac{1}{2^4} > 1.$$

Thus there is no instantaneous binary code for S with word-lengths $1, 2, 3, 3, 4$. There is, however, an instantaneous ternary code with these word-lengths: if $r = 3$ then choices of $w_1, \ldots, w_5$ eliminate proportions $1/3$, $1/3^2$, $1/3^3$, $1/3^3$, $1/3^4$ of the ternary tree T^*, where $|T| = 3$. Since

$$\frac{1}{3} + \frac{1}{3^2} + \frac{1}{3^3} + \frac{1}{3^3} + \frac{1}{3^4} \le 1,$$

such choices are possible.

Exercise 1.11

Find an instantaneous ternary code with word-lengths $1, 2, 3, 3, 4$. Is there one with word-lengths $1, 1, 2, 2, 2, 2$?

This concept of the "proportion" of an infinite tree T^* is useful, but imprecise. By making it more precise, we can use arguments like those above to give necessary and sufficient conditions for the existence of instantaneous r-ary codes with given word-lengths.

1.5 Kraft's Inequality

Motivated by the examples in §1.4, we have the following result, known as *Kraft's inequality* [Kr49]:

Theorem 1.20

There is an instantaneous r-ary code C with word-lengths $l_1, \ldots . l_q$ if and only if

$$\sum_{i=1}^{q} \frac{1}{r^{l_i}} \leq 1. \tag{1.5}$$

Proof

($\Leftarrow$) Renumbering the word-lengths if necessary, we may assume that $l_1 \leq \cdots \leq l_q$. Let $l = \max(l_1, \ldots, l_q)$, and consider the part $T^{\leq l} = T^0 \cup T^1 \cup \cdots \cup T^l$ of T^* up to height l. This is a finite tree: at each height $h = 0, 1, \ldots, l$ it has r^h vertices, the elements of T^h, or words of length h; its "leaves" are the r^l vertices at maximum height l.

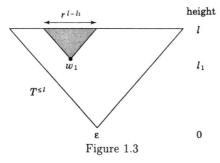

Figure 1.3

Let us assign any vertex w_1 at height l_1 to s_1, giving a code-word of length l_1, and then prune (delete) w_1 and all of $T^{\leq l}$ above w_1, since we cannot use such vertices again; in particular, this removes r^{l-l_1} leaves, the words of length l beginning with w_1 (see Figure 1.3). If $q > 1$ then

$$r^{l-l_1} < r^l \sum_{i=1}^{q} \frac{1}{r^{l_i}} \leq r^l,$$

by (1.5), so at least one leaf of $T^{\leq l}$ remains unpruned. The first l_2 symbols of this leaf give us a code-word w_2 of length l_2, which is not above w_1 in the tree. We now prune w_2 and all of $T^{\leq l}$ above it, thus removing a further r^{l-l_2} leaves (no leaf can be above both w_1 and w_2). We continue like this, choosing code-words and then pruning, so that no code-word is above or below any other. After k code-words $w_1, \ldots, w_k$ have been chosen, where $k < q$, we have pruned $r^{l-l_1} + \cdots + r^{l-l_k}$ of the r^l leaves; since

$$\sum_{i=1}^{k} r^{l-l_i} < r^l \sum_{i=1}^{q} \frac{1}{r^{l_i}} \leq r^l,$$

at least one leaf remains, so it is possible to choose w_{k+1} of length l_{k+1} as a prefix of such a leaf. We can therefore continue choosing code-words until w_q has been chosen. Throughout this process, the prefix condition is satisfied, so the resulting code $\mathcal{C} = \{w_1, \ldots, w_q\}$ is instantaneous by Theorem 1.17.

($\Rightarrow$) If $\mathcal{C}$ is instantaneous then it is a prefix code (by Theorem 1.17), so every leaf of $T^{\leq l}$ is above at most one code-word of $\mathcal{C}$. Each code-word w_i of $\mathcal{C}$ is below r^{l-l_i} leaves (where $l_i = |w_i|$), so summing over all w_i in $\mathcal{C}$ we find that there are $\sum_i r^{l-l_i}$ leaves above code-words. Since there are r^l leaves in $T^{\leq l}$, we must have

$$\sum_{i=1}^{q} r^{l-l_i} \leq r^l,$$

so dividing by r^l we get (1.5). $\square$

For illustrative examples of Kraft's inequality, see §1.4.

1.6 McMillan's Inequality

We have seen that the class of uniquely decodable codes is strictly larger than the class of instantaneous codes, so one might expect a weaker necessary and sufficient condition than (1.5) for the existence of a uniquely decodable r-ary code with specified word-lengths. Surprisingly, this is not the case. *McMillan's inequality* [McM56] is as follows:

Theorem 1.21

There is a uniquely decodable r-ary code $\mathcal{C}$ with word-lengths $l_1, \ldots, l_q$ if and only if

$$\sum_{i=1}^{q} \frac{1}{r^{l_i}} \leq 1. \tag{1.6}$$

Proof

($\Leftarrow$) By Theorem 1.20 there is an instantaneous r-ary code with word-lengths l_i; being instantaneous, this code is uniquely decodable.

($\Rightarrow$) Let $\mathcal{C}$ be a uniquely decodable r-ary code with word-lengths $l_1, \ldots, l_q$; define

$$K = \sum_{i=1}^{q} \frac{1}{r^{l_i}}, \tag{1.7}$$

and let

$$l = \max(l_1, \ldots, l_q),$$
$$m = \min(l_1, \ldots, l_q).$$

Now consider the expansion of

$$K^n = \left(\sum_{i=1}^{q} \frac{1}{r^{l_i}} \right)^n,$$

where $n \geq 1$. This is a sum of terms

$$\frac{1}{r^{l_{i_1}}} \times \frac{1}{r^{l_{i_2}}} \times \cdots \times \frac{1}{r^{l_{i_n}}} = \frac{1}{r^j}$$

where

$$j = l_{i_1} + \cdots + l_{i_n}. \tag{1.8}$$

Here $mn \leq j \leq ln$ since $m \leq l_i \leq l$ for all i, so collecting terms together we can write

$$K^n = \sum_{j=mn}^{ln} \frac{N_{j,n}}{r^j}.$$

The coefficient $N_{j,n}$ of $1/r^j$ is the number of ways of writing j in the form (1.8) as a sum of n word-lengths (possibly with repetitions); equivalently, $N_{j,n}$ is the number of sequences $w_{i_1}, \ldots, w_{i_n}$ of n code-words of $\mathcal{C}$, of total length j. Each such sequence determines a code-sequence $\mathbf{t} = w_{i_1} \ldots w_{i_n}$ of length j, and since $\mathcal{C}$ is uniquely decodable, each $\mathbf{t}$ arises from at most one such sequence of code-words. Thus $N_{j,n}$ is at most equal to the number of code-sequences $\mathbf{t}$ of length j, that is, $N_{j,n} \leq r^j$. Since K^n is a sum of $ln+1-mn$ terms $N_{j,n}/r^j \leq 1$, we have

$$K^n \leq (l-m)n + 1 \tag{1.9}$$

for all $n \geq 1$. Now K, l and m are independent of n, so if $K > 1$ then the left-hand side in this inequality grows exponentially, while the right-hand side grows only linearly. This contradicts (1.9) for sufficiently large n, so we must have $K \leq 1$.   □

The above proof is due to Karush [Ka61]; the original proof used complex functions (see [McM56] or [Re61, pp. 147–8]). Theorems 1.20 and 1.21 immediately imply:

Corollary 1.22

There is an instantaneous r-ary code with word-lengths $l_1, \ldots, l_q$ if and only if there is a uniquely decodable r-ary code with these word-lengths.

1.7 Comments on Kraft's and McMillan's Inequalities

Comment 1.23

Theorems 1.20 and 1.21 do *not* say that an r-ary code with word-lengths $l_1, \ldots, l_q$ is instantaneous or uniquely decodable if and only if $\sum r^{-l_i} \leq 1$. For instance, the binary code $\mathcal{C} = \{0, 01, 011\}$ has $l_i = 1, 2, 3$, so $\sum r^{-l_i} = \frac{7}{8} \leq 1$; however, $\mathcal{C}$ is not a prefix code, so it is not instantaneous. Similarly, one can find a binary code with these word-lengths, which is not uniquely decodable: $\{0, 01, 001\}$ is an obvious example. However:

Comment 1.24

Theorems 1.20 and 1.21 assert that if $\sum r^{-l_i} \leq 1$, then there exist codes with these parameters which *are* instantaneous and uniquely decodable: for instance, the binary code $\{0, 10, 110\}$ is a prefix code, and hence satisfies both conditions.

Comment 1.25

If an r-ary code $\mathcal{C}$ is uniquely decodable, then it need not be instantaneous, but by Corollary 1.22 there must be an instantaneous r-ary code with the same word-lengths. For instance the binary code $\mathcal{C} = \{0, 01, 11\}$ in Example 1.14 is uniquely decodable but not instantaneous; with the same word-lengths we have the instantaneous code $\mathcal{D} = \{0, 10, 11\}$ in Example 1.16.

Comment 1.26

The summand r^{-l_i} in $K = \sum r^{-l_i}$ corresponds to the rather imprecise notion of the "proportion" of the tree T^* above a vertex w_i of height l_i, as used in §1.4. This interpretation helps to explain why we need $K \leq 1$.

1.8 Supplementary Exercises

Exercise 1.12

Is there an instantaneous code over $\mathbf{Z}_3$ with word-lengths $l_i = 1, 2, 2, 2, 2, 2, 3, 3, 3, 3$? Construct one with $l_i = 1, 2, 2, 2, 2, 2, 3, 3, 3$; how many such codes are there?

Exercise 1.13

The binary code $C = \{0, 10, 11\}$ is used; how many code-sequences of length j can be formed from C ? (Hint: find and solve a recurrence relation for this number N_j.)

Exercise 1.14

Suppose that $|T| = r$, $1 \leq l_1 \leq \cdots \leq l_q$, and $\sum r^{-l_i} \leq 1$. In how many ways can one choose words $w_1, \ldots, w_q \in T^*$ so that $|w_i| = l_i$ and $\{w_1, \ldots, w_q\}$ is an instantaneous code?

Exercise 1.15

A code is *exhaustive* if every sufficiently long sequence of code-symbols begins with a code-word (equivalently, every infinite sequence of code-symbols can be decomposed into code-words). Find an equivalent condition in terms of the leaves of the tree $T^{\leq l}$ in the proof of Theorem 1.20. Which of the codes in the examples in this chapter are exhaustive?

Exercise 1.16

Show that if an r-ary code C with word-lengths $l_1, \ldots, l_q$ is exhaustive, then $\sum r^{-l_i} \geq 1$, with equality if and only if C is instantaneous.

Exercise 1.17

Let C be an r-ary code with word-lengths $l_1, \ldots, l_q$. Show that any two of the following conditions imply the third:

(a) $\mathcal{C}$ is instantaneous;

(b) $\mathcal{C}$ is exhaustive;

(c) $\sum r^{-l_i} = 1$.

Show that no one of these conditions implies any other.

2
Optimal Codes

Men of few words are the best men. (*King Henry V*)

We saw in Chapter 1 how to encode information so that decoding is unique or instantaneous. In either case the basic requirement, given by Kraft's or McMillan's inequality, is that we should use sufficiently long code-words. This raises the question of efficiency: if the code-words are too long, then storage may be difficult and transmission may be slow. We therefore need to strike a balance between using words which are long enough to allow effective decoding, and short enough for economy. From this point of view, the best codes available are those called optimal codes, the instantaneous codes with least average word-length. We will prove that they exist, and we will examine Huffman's algorithm for constructing them. For simplicity, we will concentrate mainly on the binary case ($r = 2$), though we will briefly outline how these ideas extend to non-binary codes.

2.1 Optimality

Let S be a source, as in Chapter 1. We assume, as before, that the probabilities

$$p_i = \Pr(X_n = s_i) = \Pr(s_i)$$

are independent of time n and of previous symbols $X_1, \ldots, X_{n-1}$. The following theory can be extended to apply to sources which do not satisfy these conditions, but we will concentrate on the simplest case, where these conditions hold. Since the numbers p_i form a probability distribution, we have

$$0 \leq p_i \leq 1, \qquad \sum_{i=1}^{q} p_i = 1. \tag{2.1}$$

If a code $\mathcal{C}$ for S has word-lengths $l_1, \ldots, l_q$, then its average word-length is

$$L = L(\mathcal{C}) = \sum_{i=1}^{q} p_i l_i \, . \tag{2.2}$$

Clearly $L(\mathcal{C}) > 0$ for all codes $\mathcal{C}$. For economy and efficiency, we try to make $L(\mathcal{C})$ as small as possible, while retaining instantaneous decoding: given r and the probability distribution (p_i), we try to find instantaneous r-ary codes $\mathcal{C}$ minimising $L(\mathcal{C})$. Such codes are called *optimal* or *compact* codes.

Example 2.1

Let S be the daily weather (as in Example 1.2), with $p_i = \frac{1}{4}, \frac{1}{2}, \frac{1}{4}$ for $i = 1, 2, 3$. The binary code $\mathcal{C} : s_1 \mapsto 00, s_2 \mapsto 01, s_3 \mapsto 1$ is instantaneous (because it is a prefix code), and it has

$$L(\mathcal{C}) = \frac{1}{4} \cdot 2 + \frac{1}{2} \cdot 2 + \frac{1}{4} \cdot 1 = 1.75 \, ;$$

however, the binary code $\mathcal{D} : s_1 \mapsto 00, s_2 \mapsto 1, s_3 \mapsto 01$ (which uses the same code-words but in a different order) is also instantaneous but has

$$L(\mathcal{D}) = \frac{1}{4} \cdot 2 + \frac{1}{2} \cdot 1 + \frac{1}{4} \cdot 2 = 1.5 \, .$$

Thus $L(\mathcal{D}) < L(\mathcal{C})$, and it is not hard to see that $\mathcal{D}$ is an optimal binary code for S, that is, $L(\mathcal{D}) \leq L(\mathcal{C})$ for *every* instantaneous binary code $\mathcal{C}$ for S.

This example illustrates the general rule that the average word-length is reduced by assigning shorter code-words to more frequent source-symbols (using $w_2 = 1$ in $\mathcal{D}$, rather than $w_2 = 01$ in $\mathcal{C}$); Morse code uses the same idea.

Exercise 2.1

Show that in any optimal code, if $p_i > p_j$ then $l_i \leq l_j$.

We will use this principle more systematically in later sections, to construct optimal codes for arbitrary sources. First we show that one cannot reduce the average word-length further by allowing uniquely decodable (rather than instantaneous) codes, so there is no loss in restricting to instantaneous codes. This follows immediately from Corollary 1.22:

Lemma 2.2

Given a source S and an integer r, the set of all average word-lengths $L(C)$ of uniquely decodable r-ary codes C for S is equal to the set of all average word-lengths $L(C)$ of instantaneous r-ary codes C for S.

This set of average word-lengths is clearly bounded below (by 0), so let us write $L_{\min}(S)$ for its greatest lower bound (the value of r being understood). An instantaneous r-ary code C is defined to be *optimal* if $L(C) = L_{\min}(S)$. It is not entirely obvious that such codes exist: conceivably, the instantaneous r-ary codes for S might have average word-lengths which approach but do not attain their greatest lower bound (like the numbers $1/n$ for $n \in \mathbf{N}$, with greatest lower bound 0). We therefore need to prove that optimal codes always exist (a point which is generally overlooked in the literature).

Theorem 2.3

Each source S has an optimal r-ary code for each integer $r \geq 2$.

Proof

By renumbering the source-symbols $s_1, \ldots, s_q$ if necessary, we can assume that there is some k such that $p_i > 0$ for $i \leq k$, and $p_i = 0$ for $i > k$. Let $p = \min(p_1, \ldots, p_k)$, so $p > 0$.

There certainly exists an instantaneous r-ary code C for S: for instance, we can put $l_1 = \cdots = l_q = l$ for some l such that $r^l \geq q$, and apply Theorem 1.20. To prove the theorem, it is sufficient to show that among all instantaneous r-ary codes $\mathcal{D}$ for S, there are only finitely many values $L(\mathcal{D}) \leq L(C)$; the least of these finitely many values is attained by some code $\mathcal{D}$, which must then be optimal.

To show this, let $\mathcal{D}$ be any instantaneous r-ary code with $L(\mathcal{D}) \leq L(C)$. Then the word-lengths $l_1, \ldots, l_q$ of $\mathcal{D}$ must satisfy

$$l_i \leq \frac{L(C)}{p} \quad \text{for} \quad i = 1, \ldots, k,$$

for otherwise we would have

$$L(\mathcal{D}) = p_1 l_1 + \cdots + p_q l_q \geq p_i l_i > p\frac{L(C)}{p} = L(C).$$

There are only finitely many words $w \in T^+$ with $|w| \leq L(C)/p$, so there are only finitely many choices for the code-words $w_1, \ldots, w_k$ in $\mathcal{D}$. There are infinitely many choices for each w_i with $i > k$, but they have no effect on $L(\mathcal{D})$ since $p_i = 0$ for such i. Consequently, there are only finitely many possible values of $L(\mathcal{D}) \leq L(C)$.   □

Exercise 2.2

Interpret the proof of Theorem 2.3 geometrically, the problem being to minimise the scalar product $L = \sum p_i l_i = \mathbf{p}.\mathbf{l}$, where the vectors $\mathbf{p} = (p_1, \ldots, p_q)$ and $\mathbf{l} = (l_1, \ldots, l_q)$ in $\mathbf{R}^q$ are subject to certain restrictions. (Hint: it may help to consider the case $q = 2$ first.)

2.2 Binary Huffman Codes

In 1952, Huffman [Hu52] introduced an algorithm for constructing optimal codes. For simplicity we will concentrate on the binary case, so let $T = \mathbf{Z}_2 = \{0, 1\}$. Given a source $\mathcal{S}$, we renumber the source-symbols $s_1, \ldots, s_q$ so that

$$p_1 \geq p_2 \geq \cdots \geq p_q .$$

We form a *reduced source* $\mathcal{S}'$ by amalgamating the two least-likely symbols s_{q-1} and s_q into a single symbol $s' = s_{q-1} \vee s_q$ (meaning "s_{q-1} or s_q") of probability $p' = p_{q-1} + p_q$ (if the pair s_{q-1}, s_q is not unique, we make an arbitrary choice of two least-likely symbols). Thus $\mathcal{S}'$ is a source having $q - 1$ symbols $s_1, \ldots, s_{q-2}, s'$ with probabilities $p_1, \ldots, p_{q-2}, p'$.

Given any binary code $\mathcal{C}'$ for $\mathcal{S}'$, we can form a binary code $\mathcal{C}$ for $\mathcal{S}$: if $\mathcal{C}'$ encodes s_i as w_i for $i = 1, \ldots, q - 2$, then so does $\mathcal{C}$; if $\mathcal{C}'$ encodes s' as w', then $\mathcal{C}$ encodes s_{q-1} and s_q as $w'0$ and $w'1$. This process is illustrated as follows:

$$\mathcal{S}: \quad s_1, \ \ldots \ , s_{q-2}, \ \underbrace{s_{q-1}, \ s_q}$$
$$\mathcal{S}': \quad s_1, \ \ldots \ , s_{q-2}, \qquad s'$$

$$p_1, \ \ldots \ , p_{q-2}, \ \underbrace{p_{q-1}, \ p_q}$$
$$p_1, \ \ldots \ , p_{q-2}, \qquad p'$$

$$\mathcal{C}: \quad w_1, \ \ldots \ , w_{q-2}, \ \underbrace{w'0, \ w'1}$$
$$\mathcal{C}': \quad w_1, \ \ldots \ , w_{q-2}, \qquad w'$$

Lemma 2.4

If the code $\mathcal{C}'$ is instantaneous then so is $\mathcal{C}$.

Proof

It is easy to check that if C' is a prefix code then so is C; Theorem 1.17 now completes the proof. □

This means that an instantaneous binary code C for S (which has q symbols) can be obtained from an instantaneous binary code C' for S' (which has $q - 1$ symbols). Similarly, an instantaneous binary code C' for S' can be obtained from an instantaneous binary code C'' for S'' (which has $q - 2$ symbols), where S'' is the source formed from S' by amalgamating its two least likely symbols. If we continue to reduce sources in this way, we obtain a sequence of sources $S, S', \ldots, S^{(q-2)}, S^{(q-1)}$ with the number of symbols successively equal to $q, q - 1, \ldots, 2, 1$:

$$S \to S' \to \cdots \to S^{(q-2)} \to S^{(q-1)}.$$

Now $S^{(q-1)}$ has a single symbol $s_1 \vee \cdots \vee s_q$ of probability 1, and we use the empty word ε to encode this, giving a code[1] $C^{(q-1)} = \{\varepsilon\}$ for $S^{(q-1)}$. The above process of adding 0 and 1 to a code-word w' then gives us an instantaneous binary code $C^{(q-2)} = \{\varepsilon 0 = 0, \varepsilon 1 = 1\}$ for $S^{(q-2)}$, and by repeating this process $q - 1$ times we get a sequence of binary codes $C^{(q-1)}, C^{(q-2)}, \ldots, C', C$ for the sources $S^{(q-1)}, S^{(q-2)}, \ldots, S', S$:

$$S \to S' \to \cdots \to S^{(q-2)} \to S^{(q-1)}$$
$$C \leftarrow C' \leftarrow \cdots \leftarrow C^{(q-2)} \leftarrow C^{(q-1)}.$$

The final code C is known as a *Huffman code* for S. It is instantaneous by repeated use of Lemma 2.4, and we will show that it is optimal in §2.4. (Notice that each $C^{(i)}$ is a Huffman code for $S^{(i)}$, since we can choose to ignore $S^{(j)}$ and $C^{(j)}$ for all $j < i$.)

Example 2.5

Let S have $q = 5$ symbols $s_1, \ldots, s_5$, with probabilities $p_i = 0.3, 0.2, 0.2, 0.2, 0.1$. We first perform the sequence of source-reductions; the successive probability distributions are as follows:

S :			0.3	0.2	0.2	0.2	0.1
S' :			0.3	**0.3**	0.2	0.2
S'' :		**0.4**	0.3	0.3
S''' :	**0.6**	0.4
$S^{(4)}$:	1

---

[1] Strictly speaking, $C^{(q-1)}$ is not a code, since it contains the empty word; however, the point is that it allows us to create $C^{(q-2)}, \ldots, C$, which really are codes.

Given any row, the next row is formed by replacing the two smallest probabilities with their sum (shown in bold type), positioned so that the new set of
probabilities are in non-increasing order. We now reverse this process to construct the codes for these sources, starting at the bottom with ε, and working
upwards in the reverse pattern:

$\mathcal{C}$:				00		10	11	010	011
$\mathcal{C}'$:				00	**01**	10	11
$\mathcal{C}''$:			**1**	00	01
$\mathcal{C}'''$:		**0**	1
$\mathcal{C}^{(4)}$:	ε

Each row is constructed from the row below it by taking the code-word w' for
the new symbol in that lower row (shown in bold type), and adding a final 0
or 1 to obtain code-words $w'0, w'1$ for the two amalgamated symbols; all other
code-words in the lower row are carried forward unchanged. We finish with a
binary code $\mathcal{C} = \{00, 10, 11, 010, 011\}$ for $\mathcal{S}$. It is clearly a prefix code, so it is
instantaneous. It has word-lengths $l_i = 2, 2, 2, 3, 3$, so

$$L(\mathcal{C}) = \sum p_i l_i = 0.6 + 0.4 + 0.4 + 0.6 + 0.3 = 2.3 \,.$$

In most cases, the reduction process is unique, and hence the Huffman code
$\mathcal{C}$ is unique, apart from an arbitrary decision at each stage whether to assign
code-words $w'0, w'1$ or $w'1, w'0$ respectively to the two least likely symbols;
by convention, one usually chooses the former option, but this choice has no
effect on the word-lengths. In other cases, however, there may at some stage
be more than one least likely pair, so the reduction process may not be unique,
giving a wider choice of Huffman codes $\mathcal{C}$; this happens at stage one in the
above example, and Exercise 2.3 also illustrates this behaviour. However, the
optimality of Huffman codes (to be proved later) implies that in such cases
$L(\mathcal{C})$ is nevertheless unique.

Exercise 2.3

Construct a binary Huffman code for a source with probabilities

$$p_i = 0.4, \, 0.3, \, 0.1, \, 0.1, \, 0.06, \, 0.04 \,,$$

and find its average word-length. To what extent are the code, the word-
lengths, and the average word-length unique?

The construction of Huffman codes implies that each word-length l_i is equal
to the number of times the corresponding symbol s_i of $\mathcal{S}$ is amalgamated during the reduction process. This is because the construction of w_i starts with

ε, of length 0, and adds a final symbol 0 or 1 precisely when s_i is amalgamated. Thus symbols with high probabilities, being amalgamated infrequently, are assigned short code-words, which helps to explain why Huffman codes are optimal (though it does not prove it).

To see how the probability distribution affects the construction of Huffman codes, we now look at another example.

Example 2.6

Let S have $q = 5$ symbols $s_1, \ldots, s_5$ again, but now suppose that they are equiprobable, that is, $p_1 = \cdots = p_5 = 0.2$. The reduction and encoding processes now look like this:

S :					0.2	0.2	0.2	0.2	0.2
S' :				**0.4**	0.2	0.2	0.2
S'' :			**0.4**	0.4	0.2
S''' :		**0.6**	0.4
$S^{(4)}$:	**1**

C :					01	10	11	000	001
C' :				**00**	01	10	11
C'' :			**1**	00	01
C''' :		**0**	1
$C^{(4)}$:	ε

This gives a Huffman code $\mathcal{C} = \{01, 10, 11, 000, 001\}$ for S; it is a prefix code, and is therefore instantaneous. Its word-lengths are $l_i = 2, 2, 2, 3, 3$, so its average word-length is

$$L(\mathcal{C}) = \frac{1}{5}(2 + 2 + 2 + 3 + 3) = 2.4.$$

This is slightly greater than the value 2.3 achieved in Example 2.5, where the symbols s_i were not equiprobable.

In general, the greater the variation among the probabilities p_i, the lower the average word-length of an optimal code, because there is greater scope for assigning shorter code-words to more frequent symbols. We will study this phenomenon more systematically in Chapter 3, using a concept called entropy to measure the amount of variation in a probability distribution.

Exercise 2.4

A source has three symbols, with probabilities $p_1 \geq p_2 \geq p_3$; show that a binary Huffman code for this source has average word-length $2 - p_1$.

What is the corresponding result for a source with four symbols, with probabilities $p_1 \geq p_2 \geq p_3 \geq p_4$?

2.3 Average Word-length of Huffman Codes

Let us go back to the general situation in §2.2, and compare the average word-lengths of the codes C and C'. In C', the symbol $s' = s_{q-1} \vee s_q$ has probability $p' = p_{q-1} + p_q$, and is assigned a code-word w'; let l denote the word-length $|w'|$. In C, the symbol s' is replaced with the symbols s_{q-1} and s_q of probabilities p_{q-1} and p_q, and these are assigned the code-words $w'0$ and $w'1$ of length $l+1$. All other symbols $s_1, \ldots, s_{q-2}$ are assigned the same code-words in C' as they are in C, so

$$L(C) - L(C') = p_{q-1}(l+1) + p_q(l+1) - (p_{q-1} + p_q)l$$
$$= p_{q-1} + p_q$$
$$= p', \tag{2.3}$$

which is the "new" probability created by reducing S to S'. If we iterate this, using the fact that $L(C^{(q-1)}) = |\varepsilon| = 0$, we find that

$$L(C) = \left(L(C) - L(C')\right) + \left(L(C') - L(C'')\right) + \cdots$$
$$\cdots + \left(L(C^{(q-2)}) - L(C^{(q-1)})\right) + L(C^{(q-1)})$$
$$= \left(L(C) - L(C')\right) + \left(L(C') - L(C'')\right) + \cdots + \left(L(C^{(q-2)}) - L(C^{(q-1)})\right)$$
$$= p' + p'' + \cdots + p^{(q-1)}, \tag{2.4}$$

the sum of all the new probabilities $p', p'', \ldots, p^{(q-1)}$ created in reducing S to $S^{(q-1)}$. For instance, in Example 2.5 (in §2.2) we add up the probabilities in bold type to get

$$L(C) = 0.3 + 0.4 + 0.6 + 1 = 2.3,$$

while in Example 2.6 we have

$$L(C) = 0.4 + 0.4 + 0.6 + 1 = 2.4.$$

This method is a great labour-saving device, since it allows us to work out $L(C)$ without having to construct C: for instance, in Example 2.5 it is clear from the probabilities $p_i = 0.3, 0.2, 0.2, 0.2, 0.1$ that by successively merging smallest pairs we must have

$$p' = 0.2 + 0.1 = 0.3,$$
$$p'' = 0.2 + 0.2 = 0.4,$$
$$p''' = 0.3 + 0.3 = 0.6,$$
$$p'''' = 0.4 + 0.6 = 1,$$

so $L(\mathcal{C}) = 0.3 + 0.4 + 0.6 + 1 = 2.3$.

Exercise 2.5

Use this method to verify your values for the average word-lengths of the codes in Exercises 2.3 and 2.4.

2.4 Optimality of Binary Huffman Codes

In this section we will prove that binary Huffman codes are optimal. First we need a definition and a lemma. We define two binary words w_1 and w_2 to be *siblings* if they have the form $x0, x1$ (or vice versa) for some word $x \in T^*$.

Lemma 2.7

Every source S has an optimal binary code $\mathcal{D}$ in which two of the longest code-words are siblings.

Proof

By Theorem 2.3, there is an optimal binary code for S; among all such codes, let us choose a code $\mathcal{D}$ which minimises $\sigma(\mathcal{D}) = \sum_i l_i$, the sum of the word-lengths for $\mathcal{D}$ (this is possible because word-lengths are non-negative integers). We claim that $\mathcal{D}$ has the required property.

Choose a longest code-word in $\mathcal{D}$; this must have the form xt where $x \in T^*$ and $t \in T = \mathbf{Z}_2$. Let $\bar{t}$ denote $1 - t$, so $\bar{t} = 0$ or 1 as $t = 1$ or 0 respectively. If $x\bar{t} \in \mathcal{D}$ then $xt, x\bar{t}$ are the required siblings, and we are home, so assume that $x\bar{t} \notin \mathcal{D}$. Being instantaneous, $\mathcal{D}$ is a prefix code. Now the only code-word in $\mathcal{D}$ with prefix x is xt (since $|xt|$ is maximal and $x\bar{t} \notin \mathcal{D}$), so if we replace xt in $\mathcal{D}$ with x, we get a new code $\mathcal{D}'$ for S which is also a prefix code. Thus $\mathcal{D}'$ is an instantaneous code for S with $L(\mathcal{D}') \leq L(\mathcal{D})$ and $\sigma(\mathcal{D}') = \sigma(\mathcal{D}) - 1 < \sigma(\mathcal{D})$, against our choice of $\mathcal{D}$. Thus $x\bar{t} \in \mathcal{D}$, as required.   □

Theorem 2.8

If $\mathcal{C}$ is a binary Huffman code for a source S, then $\mathcal{C}$ is an optimal code for S.

Proof

Lemma 2.4 shows that $\mathcal{C}$ is instantaneous, so it is sufficient to show that $L(\mathcal{C})$ is minimal (among the average word-lengths of all instantaneous binary codes for S). We use induction on the number q of source-symbols. If $q = 1$ then $\mathcal{C} = \{\varepsilon\}$ with $L(\mathcal{C}) = 0$, so the result is trivially true. (Strictly, $\{\varepsilon\}$ is not a code, but see the footnote in §2.2.) Hence we may assume that $q > 1$, and that the result is proved for all sources with $q - 1$ symbols.

Let S' be the source obtained by reducing S as in §2.2, so S' has $q-1$ symbols $s_1, \ldots, s_{q-2}, s' = s_{q-1} \vee s_q$. By (2.3) we have $L(C) - L(C') = p_{q-1} + p_q = p'$, the probability of s'.

Now let $D : s_i \mapsto x_i$ be the optimal binary code for S given by Lemma 2.7, so D has a sibling pair of longest code-words $x_u = x0$, $x_v = x1$, representing symbols s_u, s_v of S; we will show that we can assume that $u = q - 1$ and $v = q$.

If $v \neq q$ then we can transpose the code-words x_v and x_q assigned to s_v and s_q, giving another instantaneous code D^* for S; if m_i denotes $|x_i|$ then this transposition replaces the terms $p_v m_v + p_q m_q$ in $L(D)$ with $p_v m_q + p_q m_v$ in $L(D^*)$. Now

$$(p_v m_v + p_q m_q) - (p_v m_q + p_q m_v) = (p_v - p_q)(m_v - m_q) \geq 0$$

since $p_v \geq p_q$ and $m_v \geq m_q$, so $L(D) \geq L(D^*)$. Since D is optimal, this implies that $L(D) = L(D^*)$ and D^* is optimal. Replacing D with D^* if necessary, we may therefore assume that $v = q$. A similar argument allows us to assume that $u = q - 1$, so the siblings $x0$ and $x1$ in D are the code-words for s_{q-1} and s_q.

We now form a code D' for S', given by $s_i \mapsto x_i$ for $i = 1, \ldots, q - 2$, and $s' \mapsto x$. Thus the relationship of D to D' is the same as the relationship of C to C'. In particular, the argument in §2.3, applied to D and D', shows that

$$L(D) - L(D') = p_{q-1} + p_q = L(C) - L(C'),$$

so

$$L(D') - L(C') = L(D) - L(C).$$

Now C' is a Huffman code for S', a source with $q-1$ symbols, so by the induction hypothesis C' is optimal; thus $L(C') \leq L(D')$ and hence $L(C) \leq L(D)$. Since D is optimal, so is C (with $L(C) = L(D)$). □

Exercise 2.6

Comment on the following argument: every source has a Huffman code; all Huffman codes are optimal; hence every source has an optimal code.

2.5 r-ary Huffman Codes

If we use an alphabet T with $|T| = r > 2$, then the construction of r-ary Huffman codes is similar to that in the binary case. Given a source S, we form a sequence of reduced sources $S', S'', \ldots$, each time amalgamating the r least likely symbols s_i into a single symbol s', and adding their probabilities to give the probability p' of s'.

We want eventually to reduce S to a source with a single symbol (of probability 1), which is assigned the code-word ε. Since each step of the reduction process reduces the number of symbols by $r - 1$, this is possible if and only if $q \equiv 1 \bmod (r - 1)$. This condition is always satisfied when $r = 2$, but not necessarily when $r > 2$. If $q \not\equiv 1 \bmod (r - 1)$, we adjoin enough extra symbols s_i to S, with probabilities $p_i = 0$, thus increasing q so that the congruence is satisfied, and then carry out the reduction process.

Example 2.9

Let $q = 6$ and $r = 3$. Since $r - 1 = 2$ we need $q \equiv 1 \bmod (2)$, so we adjoin an extra symbol s_7 to S, with $p_7 = 0$. The reduction process now gives sources S', S'' and S''' with the number of symbols equal to $5, 3$ and 1.

The construction of the code C is similar to that in the binary case. Given a code $C^{(i)}$ for $S^{(i)}$, we form a code $C^{(i-1)}$ for $S^{(i-1)}$: this is done by removing the code-word w' for the new symbol s' of $S^{(i)}$, and replacing it with r code-words $w't$ $(t \in T)$ for the r symbols of $S^{(i-1)}$ which were amalgamated to form s'. By iterating this process, we eventually get an r-ary Huffman code C for S, deleting the code-words for any extra symbols s_i which were adjoined at the beginning.

Example 2.10

Let $q = 6$ and $r = 3$ as in Example 2.9, and suppose that the symbols $s_1, \ldots, s_6$ of S have probabilities $p_i = 0.3, 0.2, 0.2, 0.1, 0.1, 0.1$. After adjoining s_7 with $p_7 = 0$, we find that the reduction process is as follows:

S :			0.3	0.2	0.2		0.1	0.1	0.1	0
S' :			0.3	0.2	0.2	**0.2**	0.1	
S'' :		**0.5**	0.3	0.2			
S''' :		**1**					

If we take $T = \mathbf{Z}_3 = \{0, 1, 2\}$, then one possible encoding process is:

C :			1	2	00		02	010	011	012
C' :			1	2	00	**01**	02	
C'' :		**0**	1	2			
C''' :		ε					

Deleting the code-word 012 for the adjoined symbol s_7, we obtain a ternary Huffman code $C = \{1, 2, 00, 02, 010, 011\}$ for S, with $L(C) = 1.7$.

The proof that r-ary Huffman codes are instantaneous is similar to that for $r = 2$; however, the proof of optimality is a little harder than in the binary case, since Lemma 2.7 does not extend quite so easily to the case $r > 2$, so we will omit it. The proof of (2.4), that $L(\mathcal{C})$ is the sum $p' + p'' + \cdots$ of all the "new" probabilities, also applies in the non-binary case: for instance, $L(\mathcal{C}) = 0.2 + 0.5 + 1 = 1.7$ in Example 2.10.

Exercise 2.7

Find binary and ternary Huffman codes for a source with probabilities

$$p_i = 0.3,\, 0.2,\, 0.15,\, 0.1,\, 0.1,\, 0.08,\, 0.05,\, 0.02\,.$$

Find the average word-length in each case.

Exercise 2.8

Extend the proof of (2.4) to r-ary codes for $r > 2$.

2.6 Extensions of Sources

Instead of encoding source symbols s_i one at a time, it can be more efficient to encode blocks of consecutive symbols, for instance words (or even sentences) of a text, rather than individual letters. This gives more variation of probabilities, and hence allows lower average word-lengths (as noted in §2.2).

Let $\mathcal{S}$ be a source with a source alphabet S of q symbols $s_1, \ldots, s_q$, of probabilities $p_1, \ldots, p_q$. The n-th *extension* $\mathcal{S}^n$ of $\mathcal{S}$ is the source with source alphabet S^n consisting of the q^n symbols $s_{i_1} \ldots s_{i_n}$ $(s_{i_j} \in S)$, of probabilities $p_{i_1} \ldots p_{i_n}$. We can think of a symbol $s_{i_1} \ldots s_{i_n}$ of $\mathcal{S}^n$ as a block of n consecutive symbols from $\mathcal{S}$, or alternatively as a single output from n independent copies of $\mathcal{S}$ emitting symbols simultaneously (imagine tossing several similar coins, or rolling several similar dice). We can check that the probabilities $p_{i_1} \ldots p_{i_n}$ form a probability distribution by expanding the left-hand side of the equation

$$(p_1 + \cdots + p_q)^n = 1^n = 1\,,$$

and noting that each $p_{i_1} \ldots p_{i_n}$ appears once.

Example 2.11

Let $\mathcal{S}$ have source alphabet $S = \{s_1, s_2\}$ with $p_1 = 2/3$, $p_2 = 1/3$. Then $\mathcal{S}^2$ has source alphabet $S^n = \{s_1 s_1,\ s_1 s_2,\ s_2 s_1,\ s_2 s_2\}$ with probabilities 4/9, 2/9, 2/9, 1/9.

In general, let p_1 and p_q be the greatest and least of the probabilities for S, so p_1^n and p_q^n are the greatest and least of the probabilities for S^n. Assuming that $p_1 > p_q$ (that is, that the probabilities p_i are not all equal to $1/q$), we have

$$\frac{p_1^n}{p_q^n} = \left(\frac{p_1}{p_q}\right)^n \to \infty \quad \text{as} \quad n \to \infty;$$

this means that S^n has greater variability of probabilities as n increases, so we might expect more efficient coding.

Example 2.12

If S is as in Example 2.11, there is a binary Huffman code $C : s_1 \mapsto 0,\ s_2 \mapsto 1$ with average word-length $L(C) = 1$. It is hard to believe that one can improve on this, but nevertheless, let us construct a Huffman code for S^2. We use the algorithm described in §2.2, as follows (we have not bothered to rewrite the probabilities in decreasing order in each row):

S^2 :	$\frac{4}{9}$	$\frac{2}{9}$	$\frac{2}{9}$	$\frac{1}{9}$		0	10	110	111
$(S^2)'$:	$\frac{4}{9}$	$\frac{2}{9}$	$\frac{3}{9}$			0	10	11	
$(S^2)''$:	$\frac{4}{9}$	$\frac{5}{9}$				0	1		
$(S^2)'''$:	1					ε			

This gives a Huffman code $C^2 : s_1 s_1 \mapsto 0,\ s_1 s_2 \mapsto 10,\ s_2 s_1 \mapsto 110,\ s_2 s_2 \mapsto$ 111 for S^2, with average word-length

$$L_2 = L(C^2) = \frac{3}{9} + \frac{5}{9} + 1 = \frac{17}{9}.$$

Now each code-word in C^2 represents a block of *two* symbols from S, so on average, each symbol of S requires 17/18 binary digits. Thus, as an encoding of S, C^2 has average word-length

$$\frac{L_2}{2} = \frac{17}{18} = 0.944\ldots.$$

This is less than the average word-length $L(C) = 1$ of the Huffman code C for S, so this encoding is more efficient.

Strictly speaking, what we have described is not a *code* for S, since individual symbols of S are not assigned their own code-words; nevertheless, it enables us to encode the information coming from S, so we call it an *encoding* of S. As such, it is uniquely decodable: as a code for S^2, the Huffman code C^2 is instantaneous and hence uniquely decodable; this means that we can break any code-sequence **t** into code-words in a unique way, thus determining the symbols $s_{i_1} s_{i_2}$ of S^2 encoded in **t**, and from this we obtain the individual symbols s_i of S encoded in **t**. However, this decoding is not quite instantaneous: we have to determine symbols of S in consecutive pairs, rather than one at a time, so there is a bounded delay while we wait for pairs to be completed.

Continuing this principle, one can show that a Huffman code C^3 for S^3 has average word-length $L_3 = L(C^3) = 76/27$ (Exercise 2.9); as an encoding of S it has average word-length

$$\frac{L_3}{3} = \frac{76}{81} = 0.938\ldots,$$

which is even better than using C^2.

There is an obvious extension of this idea to S^n for any n, and two natural questions arise: what happens to the average word-length L_n/n as $n \to \infty$, where $L_n = L(C^n)$, and can we apply the same method to obtain more efficient encodings of other sources? To answer these questions we need the next major topic, namely entropy.

Exercise 2.9

Let S be the source in Examples 2.11 and 2.12. Find the probability distribution for S^3, and show that a binary Huffman code C^3 for S^3 has average word-length $L_3 = L(C^3) = 76/27$.

2.7 Supplementary Exercises

Exercise 2.10

Let S be a source with probabilities 0.3, 0.3, 0.2, 0.2; how many optimal binary codes does S have? Are they all Huffman codes?

Exercise 2.11

A source S has symbols $s_1, \ldots, s_q$ with probabilities $p_1 \geq \cdots \geq p_q$ satisfying $p_i > p_{i+2} + \cdots + p_q$ for $i = 1, \ldots, q - 3$. Prove that in any binary Huffman code for S, the word-lengths are $1, 2, \ldots, q - 1, q - 1$. How many distinct binary Huffman codes are there for S? For each $q \geq 1$, give an example of a probability distribution (p_i) satisfying these inequalities.

Exercise 2.12

How can r-ary Huffman coding be implemented so as to give a Huffman code which also minimises the total word-length $\sigma(\mathcal{C}) = \sum_i l_i$? (Hint: first try binary Huffman coding, where $p_i = 1/3, 1/3, 1/6, 1/6$.)

Exercise 2.13

An unknown object s is chosen from a known finite set $S = \{s_1, \ldots, s_q\}$, each s_i having a known probability p_i of being chosen. The object has to be identified (as in the game Twenty Questions) by asking a sequence $Q_1, Q_2, \ldots$ of questions, each of which must have the form "Is s in T?" for some subset T of S. Devise a questioning strategy which minimises the average number of questions required.

Exercise 2.14

Let $\mathcal{C}$ be a binary Huffman code for a source S with q equiprobable symbols. Is it possible that $L_2/2 < L(\mathcal{C})$ in the notation of §2.6? Give some values of q such that $L_n/n = L(\mathcal{C})$ for all n.

3
Entropy

Brevity is the soul of wit. (*Hamlet*)

The aim of this chapter is to introduce the entropy function, which measures the amount of information emitted by a source. We shall examine the basic properties of this function, and show how it is related to the average word-lengths of encodings of the source.

3.1 Information and Entropy

To quantify the information conveyed by the symbols s_i of a source S, we define a number $I(s_i)$, for each i, which represents how much information is gained by knowing that S has emitted s_i; this also represents our prior uncertainty as to whether s_i will be emitted, and our surprise on learning that it has been emitted. We therefore require that:

(1) $I(s_i)$ is a decreasing function of the probability p_i of s_i, with $I(s_i) = 0$ if $p_i = 1$;

(2) $I(s_i s_j) = I(s_i) + I(s_j)$.

Condition (1) asserts that the greater the probability of an event, the less information it conveys, and an inevitable event conveys no information; newspaper editors tend to use this principle in selecting what stories to print. Condition (2) asserts that since S emits successive symbols independently (as we have been assuming), the amount of information gained by knowing two successive symbols is the sum of the two individual amounts of information. (If successive symbols were *not* independent, it would be less than the sum, since knowing s_i would tell us something about s_j).

Independence of the symbols in S means that $\Pr\left(s_i s_j\right) = \Pr\left(s_i\right)\Pr\left(s_j\right) = p_i p_j$ for all i and j. It follows that conditions (1) and (2) will be satisfied if we define

$$I(s_i) = -\log p_i = \log \frac{1}{p_i}\,, \tag{3.1}$$

so that

$$I(s_i s_j) = \log \frac{1}{p_i p_j} = \log \frac{1}{p_i} + \log \frac{1}{p_j} = I(s_i) + I(s_j).$$

Since $I(s_i) \to +\infty$ as $p_i \to 0$, we use the convention that

$$I(s_i) = +\infty \quad \text{if} \quad p_i = 0\,.$$

The graph of this function is shown in Fig. 3.1.

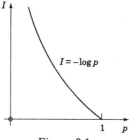

$$I = -\log p$$

Figure 3.1

The base chosen for the logarithms is not very important. We usually take $\log = \log_r$, where r is the number of code-symbols, so in the most frequent binary case we have $\log = \lg = \log_2$. A change of base of logarithms simply represents a change of units: since

$$x = r^{\log_r x}$$

for all $x > 0$, taking logarithms to another base s gives

$$\log_s x = \log_s r \cdot \log_r x\,.$$

In the binary case, the units of information are called *bits* (*binary digits*). If r is unimportant or understoood, we will write $I(s_i) = -\log(p_i)$; if we wish to emphasise the value of r, we will write $I_r(s_i) = -\log_r(p_i)$.

Example 3.1

Let S be an unbiased coin, with s_1 and s_2 representing heads and tails. Then $p_1 = p_2 = \frac{1}{2}$, so if we take $r = 2$ then $I_2(s_1) = I_2(s_2) = 1$. Thus the standard unit of information is how much we learn from a single toss of an unbiased coin.

Since each symbol s_i of a source S is emitted with probability p_i, it follows that the average amount of information conveyed by S (per source-symbol) is given by the function

$$H_r(S) = \sum_{i=1}^{q} p_i I_r(s_i) = \sum_{i=1}^{q} p_i \log_r \frac{1}{p_i} = -\sum_{i=1}^{q} p_i \log_r p_i \,,$$

called the r-ary *entropy* of S. As with the function I, a change in the base r corresponds to a change of units, given by

$$H_s(S) = \log_s r . H_r(S) \,.$$

When r is understood, or unimportant, we will simply write

$$H(S) = \sum_{i=1}^{q} p_i \log \frac{1}{p_i} = -\sum_{i=1}^{q} p_i \log p_i \,. \tag{3.2}$$

Since $p \log(1/p) = -p \log p \to 0$ as $p \to 0$ (see Fig. 3.2), we adopt the convention that $p \log(1/p) = 0$ if $p = 0$, so that $H(S)$ is a continuous function of the probabilities p_i.

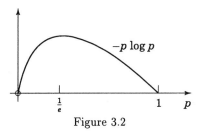

Figure 3.2

Example 3.2

Let S have $q = 2$ symbols, with probabilities p and $1 - p$; thus S could be the tossing of a coin, possibly biased. We will use this probability distribution rather frequently, so for convenience we introduce the notation

$$\bar{p} = 1 - p$$

whenever $0 \le p \le 1$. (There should be no confusion with complex conjugation, which is not used in this book.) Then

$$H(S) = -p \log p - \bar{p} \log \bar{p}.$$

We denote this important function by $H(p)$, or more precisely $H_r(p)$, so

$$H(p) = -p \log p - \bar{p} \log \bar{p}. \tag{3.3}$$

The graph of the function $H_2(p)$ is given in Fig. 3.3; for general r we simply change the vertical scale by a factor of $\log_r 2$. This shows that $H(p)$ is greatest $(= 1)$ when $p = \frac{1}{2}$, and least $(= 0)$ when $p = 0$ or 1. Thus maximum and minimum uncertainty about S correspond to maximum and minimum information conveyed by S. Note that the graph is symmetric about the vertical line $p = \frac{1}{2}$, that is,

$$H(p) = H(\bar{p}).$$

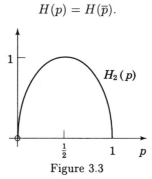

Figure 3.3

If we put $p = \frac{2}{3}$ in Example 3.2 (as in §2.6), we find that

$$H_2(S) = \frac{2}{3} \log_2 \frac{3}{2} + \frac{1}{3} \log_2 3 = \log_2 3 - \frac{2}{3} \log_2 2 = \log_2 3 - \frac{2}{3} \approx 0.918,$$

where we use the approximation $\log_2 3 \approx 1.585$; this biased coin is therefore conveying rather less information than the unbiased coin considered in Example 3.1, where $H_2(S) = 1$.

Example 3.3

If S has $q = 5$ symbols with probabilities $p_i = 0.3, 0.2, 0.2, 0.2, 0.1$, as in §2.2, Example 2.5, we find that $H_2(S) \approx 2.246$.

Example 3.4

If S has q equiprobable symbols, then $p_i = 1/q$ for each i, so

$$H_r(S) = q \cdot \frac{1}{q} \log_r q = \log_r q.$$

In particular, if we put $q = 5$, as in §2.2, Example 2.6, we find that $H_2(S) = \log_2 5 \approx 2.321$. By comparing this with Example 3.3, we see how a source with equiprobable symbols conveys more information than one with varied probabilities.

We can also compare the entropies of these sources S with the average word-lengths obtained by binary Huffman coding in Chapter 2. In Example 3.2, for instance, with $p = \frac{2}{3}$, we find that for $n = 1, 2, 3$ the average word-length obtained by binary Huffman coding of S^n in §2.6 is $L \approx 1, 0.944, 0.938$ respectively, which is approaching the entropy $H_2(S) \approx 0.918$. In Example 3.3, the average word-length $L(C) = 2.3$ obtained in Example 2.5 is close to the entropy $H_2(S) \approx 2.246$. Similarly in Example 3.4, the average word-length $L(C) = 2.4$ obtained in Example 2.6 is close to the entropy $H_2(S) \approx 2.321$. These close relationships between average word-length and entropy illustrate Shannon's First Theorem, which we will state and prove in §3.6.

Example 3.5

Putting $q = 6$ in Example 3.4, we see that an unbiased die has binary entropy $\log_2 6 \approx 2.586$.

Example 3.6.

Using the known frequencies of the letters of the alphabet, the entropy of English text has been computed as approximately 4.03.

This last example, which seems to suggest that reading a book is about four times as informative as tossing a coin, illustrates the fact that Information Theory is not concerned with how useful or interesting a message is, since these depend very much on the individual reading it. Thus a statistician might be delighted to receive a book of random numbers or letters, whereas a normal person would probably prefer a novel, even if it had lower entropy.

Exercise 3.1

A source S has probabilities $p_i = 0.3, 0.2, 0.15, 0.1, 0.1, 0.08, 0.05, 0.02$. Find $H_2(S)$ and $H_3(S)$, and compare these with the average word-lengths of binary and ternary Huffman codes for S (see Exercise 2.7).

3.2 Properties of the Entropy Function

In §3.1 we defined the entropy of a source S with probabilities p_i to be

$$H_r(S) = \sum_i p_i \log_r \frac{1}{p_i}.$$

Since $p \log_r(1/p) \geq 0$, with equality if and only if $p = 0$ or 1, we have

Theorem 3.7

$H_r(S) \geq 0$, with equality if and only if $p_i = 1$ for some i (so that $p_j = 0$ for all $j \neq i$).

Thus the entropy is least when there is no uncertainty about the symbols emitted by S, with one symbol always occurring, so that no information is conveyed. When is entropy greatest? To answer this question, we need:

Lemma 3.8

For all $x > 0$ we have $\ln x \leq x - 1$, with equality if and only if $x = 1$.

Proof

Let $f(x) = x - 1 - \ln x$, so $f(1) = 0$. Then $f'(x) = 1 - x^{-1}$ for all $x > 0$, so f has a unique stationary point, at $x = 1$. Since $f''(x) = x^{-2} > 0$ for all x, this is the unique global minimum of f, so $f(x) \geq 0$, with equality if and only if $x = 1$. $\qquad\qquad\square$

This result is illustrated in Fig. 3.4.

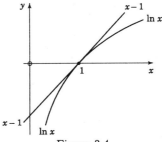

Figure 3.4

Converting this into logarithms to some other base r, we have $\log_r x \leq (x - 1) \log_r e$, with equality if and only if $x = 1$. The next result looks rather technical, but it has a number of very useful consequences.

Corollary 3.9

Let $x_i \geq 0$ and $y_i > 0$ for $i = 1, \ldots, q$, and let $\sum_i x_i = \sum_i y_i = 1$ (so (x_i) and (y_i) are probability distributions, with $y_i \neq 0$). Then

$$\sum_{i=1}^{q} x_i \log_r \frac{1}{x_i} \leq \sum_{i=1}^{q} x_i \log_r \frac{1}{y_i},$$

(that is, $\sum_i x_i \log_r (y_i/x_i) \leq 0$), with equality if and only if $x_i = y_i$ for all i.

Proof

If each $x_i \neq 0$ then the difference between the left- and right-hand sides of the inequality is

$$
\begin{aligned}
\text{LHS} - \text{RHS} &= \sum_{i=1}^{q} x_i \log_r \left(\frac{y_i}{x_i} \right) \\
&= \frac{1}{\ln r} \sum_{i=1}^{q} x_i \ln \left(\frac{y_i}{x_i} \right) && (\text{since } \log_r x = \ln x / \ln r) \\
&\leq \frac{1}{\ln r} \sum_{i=1}^{q} x_i \left(\frac{y_i}{x_i} - 1 \right) && (\text{by Lemma 3.8, used } q \text{ times}) \\
&= \frac{1}{\ln r} \left(\sum_{i=1}^{q} y_i - \sum_{i=1}^{q} x_i \right) \\
&= 0 ,
\end{aligned}
$$

with equality if and only if each $y_i/x_i = 1$. When some $x_i = 0$ the argument is similar, since our convention that $x_i \log(1/x_i) = 0$ allows us to ignore such terms. □

Theorem 3.10

If a source S has q symbols then $H_r(S) \leq \log_r q$, with equality if and only if the symbols are equiprobable.

Proof

If we put $x_i = p_i$ (the probabilities of S) and $y_i = 1/q$, then the conditions of Corollary 3.9 are satisfied. We therefore have

$$H_r(S) = \sum_{i=1}^{q} p_i \log_r \frac{1}{p_i} \leq \sum_{i=1}^{q} p_i \log_r q = \log_r q \sum_{i=1}^{q} p_i = \log_r q ,$$

with equality if and only if each $p_i = 1/q$. □

Thus the entropy is greatest, and the most information is conveyed, when there is the greatest uncertainty about the symbols emitted.

3.3 Entropy and Average Word-length

Near the end of §3.1, we considered several sources and compared their entropies with the average word-lengths of their Huffman codings. We will now explore the connection between entropy and average word-length in greater detail.

Theorem 3.11

If C is any uniquely decodable r-ary code for a source S, then $L(C) \geq H_r(S)$.

Proof

If we define

$$K = \sum_{i=1}^{q} r^{-l_i},$$

where C has word-lengths $l_1, \ldots, l_q$, then McMillan's inequality (Theorem 1.21) gives $K \leq 1$. We now use Corollary 3.9, with $x_i = p_i$ and $y_i = r^{-l_i}/K$, so $y_i > 0$ and $\sum_i y_i = 1$. Then

$$H_r(S) = \sum_{i=1}^{q} p_i \log_r \left(\frac{1}{p_i}\right)$$

$$\leq \sum_{i=1}^{q} p_i \log_r \left(\frac{1}{y_i}\right) \qquad \text{(by Corollary 3.9)}$$

$$= \sum_{i=1}^{q} p_i \log_r (r^{l_i} K)$$

$$= \sum_{i=1}^{q} p_i (l_i + \log_r K)$$

$$= \sum_{i=1}^{q} p_i l_i + \log_r K \sum_{i=1}^{q} p_i$$

$$= L(C) + \log_r K \qquad \text{(since } \sum p_i = 1\text{)}$$

$$\leq L(C)$$

(since $K \leq 1$ implies that $\log_r K \leq 0$). □

The interpretation of this is as follows: each symbol emitted by S carries $H_r(S)$ units of information, on average; if S is to be encoded without losing any of this information, then the code C must be uniquely decodable; each code-symbol conveys one unit of information, so on average each code-word of C must contain at least $H_r(S)$ code-symbols, that is, $L(C) \geq H_r(S)$. In particular, sources emitting more information require longer code-words.

Corollary 3.12

Given a source S with probabilities p_i, there is a uniquely decodable r-ary code C for S with $L(C) = H_r(S)$ if and only if $\log_r p_i$ is an integer for each i, that is, each $p_i = r^{e_i}$ for some integer $e_i \leq 0$.

Proof

($\Rightarrow$) If $L(C) = H_r(S)$ in the proof of Theorem 3.11, then both of the inequality signs there must represent equality. Thus $p_i = y_i$ for each i by Corollary 3.9, and $\log_r K = 0$. This gives $K = 1$ and $p_i = r^{-l_i}/K = r^{-l_i}$, so $\log_r p_i = -l_i$, an integer.

($\Leftarrow$) Suppose that $-\log_r p_i$ is an integer l_i for each i. Since $p_i \leq 1$ we have $l_i \geq 0$. Now $r^{l_i} = 1/p_i$ for each i, so

$$\sum_{i=1}^{q} \frac{1}{r^{l_i}} = \sum_{i=1}^{q} p_i = 1.$$

Thus McMillan's inequality (Theorem 1.21) is satisfied, so there exists a uniquely decodable r-ary code C for S with word-lengths l_i. As required, this has average word-length

$$L(C) = \sum_{i=1}^{q} p_i l_i = \sum_{i=1}^{q} p_i \log_r \frac{1}{p_i} = H_r(S).$$

$\square$

The condition $p_i = r^{e_i}$ in Corollary 3.12 is very restrictive; for most sources, every uniquely decodable code satisfies $L(C) > H_r(S)$.

Example 3.13

If S has $q = 3$ symbols s_i, with probabilities $p_i = \frac{1}{4}, \frac{1}{2}$ and $\frac{1}{4}$ (see Examples 1.2 and 2.1), then the binary entropy of S is

$$H_2(S) = \frac{1}{4} \log_2 4 + \frac{1}{2} \log_2 2 + \frac{1}{4} \log_2 4 = \frac{1}{4} \cdot 2 + \frac{1}{2} \cdot 1 + \frac{1}{4} \cdot 2 = \frac{3}{2}.$$

The code $C : s_1 \mapsto 00$, $s_2 \mapsto 1$, $s_3 \mapsto 01$ is a binary Huffman code for S, so it is optimal. It has average word-length

$$L(C) = \frac{1}{4} \cdot 2 + \frac{1}{2} \cdot 1 + \frac{1}{4} \cdot 2 = \frac{3}{2},$$

so in this case $L(C) = H_2(S)$ for some uniquely decodable binary code C. The reason for this is that the probabilities p_i are all powers of 2.

Example 3.14

Let S have $q = 5$ symbols, with probabilities $p_i = 0.3, 0.2, 0.2, 0.2, 0.1$, as in §2.2, Example 2.5. We saw in Example 3.3 that $H_2(S) \approx 2.246$, and in Example 2.5 we saw that a binary Huffman code for S has average word-length 2.3, so it follows from Theorem 2.8 that every uniquely decodable binary code C for S satisfies $L(C) \geq 2.3 > H_2(S)$. Thus no such code satisfies $L(C) = H_2(S)$, the reason being that in this case the probabilities p_i are not all powers of 2.

Exercise 3.2

For each $q \geq 2$, give an example of a source S with q symbols, and an instantaneous binary code C for S attaining the lower bound $L(C) = H_2(S)$.

By Corollary 3.12, if some $p_i = 0$ then we must have $L(C) > H_r(S)$. However, by deleting such symbols s_i we may be able to achieve equality here, reducing $L(C)$ by allowing shorter code-words, without changing the entropy.

Example 3.15

Let S have three symbols s_i with probabilities $p_i = \frac{1}{2}, \frac{1}{2}, 0$. Then $H_2(S) = 1$, but a binary Huffman code C for S has word-lengths $1, 2, 2$ with $L(C) = 1.5$. Without s_3, however, we have $H_2(S) = 1 = L(C)$, since we can now use the code $C = \{0, 1\}$; equality is possible here since the remaining non-zero probabilities p_i are powers of $r \,(= 2)$.

If C is an r-ary code for a source S, we define its *efficiency* to be

$$\eta = \frac{H_r(S)}{L(C)}, \tag{3.4}$$

so $0 \leq \eta \leq 1$ for every uniquely decodable code C for S by Theorem 3.11. The *redundancy* of C is defined to be $\bar{\eta} = 1 - \eta$; thus increasing redundancy reduces efficiency, contrary to the belief of some employers. In Examples 3.13 and 3.14, we have $\eta = 1$ and $\eta \approx 0.977$ respectively.

Exercise 3.3

A source S has probabilities $p_i = 0.4, 0.3, 0.1, 0.1, 0.06, 0.04$ (Exercise 2.3). Calculate the entropy of S, and hence find the efficiency of a binary Huffman code for S.

3.4 Shannon–Fano Coding

Huffman codes are optimal, but it can be tedious to calculate their average word-lengths. Shannon–Fano codes are close to optimal, but their average word-lengths are easier to estimate.

Let us first assume that our source S has no probabilities $p_i = 0$. By Corollary 3.12, if the average word-length $L(\mathcal{C})$ of a uniquely decodable r-ary code $\mathcal{C}$ for S is to attain the lower bound $H_r(S)$, then its word-lengths must satisfy $l_i = \log_r(1/p_i)$ for all i. This is usually impossible since the numbers $\log_r(1/p_i)$ are not generally integers. In this case, we do the next best thing and take

$$l_i = \lceil \log_r(1/p_i) \rceil, \tag{3.5}$$

where $\lceil x \rceil = \min\{n \in \mathbf{Z} \mid n \geq x\}$ denotes the least integer $n \geq x$. (This is the "ceiling function", which rounds up to the next integer.) Thus l_i is the unique integer such that

$$\log_r \frac{1}{p_i} \leq l_i < 1 + \log_r \frac{1}{p_i}, \tag{3.6}$$

and so $p_i \geq r^{-l_i}$ for each i. Summing this over all i, we find that

$$K = \sum_{i=1}^{q} r^{-l_i} \leq \sum_{i=1}^{q} p_i = 1,$$

so Theorem 1.20 (Kraft's inequality) implies that there is an instantaneous r-ary code $\mathcal{C}$ for S with these word-lengths l_i. We call $\mathcal{C}$ a *Shannon–Fano code* for S. Note that we have *not* described how to construct such codes, but have merely shown that they exist.

If we multiply (3.6) by p_i and then sum over all i, we have

$$\sum_{i=1}^{q} p_i \log_r \frac{1}{p_i} \leq \sum_{i=1}^{q} p_i l_i < \sum_{i=1}^{q} p_i \left(1 + \log_r \frac{1}{p_i} \right) = 1 + \sum_{i=1}^{q} p_i \log_r \frac{1}{p_i},$$

giving

$$H_r(S) \leq L(\mathcal{C}) < 1 + H_r(S). \tag{3.7}$$

This argument can be extended to the case where some p_i are 0, by taking limits as $p_i \to 0$ (we omit the details). However, taking limits destroys the "sharpness" of inequalities, so we now have the slightly weaker result

$$H_r(S) \leq L(\mathcal{C}) \leq 1 + H_r(S). \tag{3.8}$$

We have therefore proved:

Theorem 3.16

Every r-ary Shannon–Fano code $\mathcal{C}$ for a source $\mathcal{S}$ satisfies

$$H_r(\mathcal{S}) \leq L(\mathcal{C}) \leq 1 + H_r(\mathcal{S}).$$

Corollary 3.17

Every optimal r-ary code $\mathcal{D}$ for a source $\mathcal{S}$ satisfies

$$H_r(\mathcal{S}) \leq L(\mathcal{D}) \leq 1 + H_r(\mathcal{S}).$$

Proof

Theorem 3.11, optimality and Theorem 3.16 give $H_r(\mathcal{S}) \leq L(\mathcal{D}) \leq L(\mathcal{C}) \leq 1 + H_r(\mathcal{S})$. $\qquad\qquad\qquad\qquad\qquad\qquad\qquad\qquad\qquad\qquad\qquad\qquad\qquad$ □

This means that, even if the lower bound $H_r(\mathcal{S})$ cannot be attained, we can find codes which come reasonably close to it.

Example 3.18

Let $\mathcal{S}$ have 5 symbols, with probabilities $p_i = 0.3, 0.2, 0.2, 0.2, 0.1$ as in Example 2.5, so $1/p_i = 10/3, 5, 5, 5, 10$. A binary Shannon–Fano code $\mathcal{C}$ for $\mathcal{S}$ therefore has word-lengths

$$l_i = \lceil \log_2(1/p_i) \rceil = \min\{n \in \mathbf{Z} \mid 2^n \geq 1/p_i\} = 2, 3, 3, 3, 4$$

and hence has average word-length $L(\mathcal{C}) = \sum p_i l_i = 2.8$. Compare this with a Huffman code $\mathcal{D}$ for $\mathcal{S}$, which has $L(\mathcal{D}) = 2.3$ (see §2.2). We saw in Example 3.3 that $H_2(\mathcal{S}) \approx 2.246$, so $\mathcal{C}$ satisfies Theorem 3.16. The efficiency of $\mathcal{C}$ is $\eta \approx 2.246/2.8 \approx 0.802$, whereas $\mathcal{D}$ has $\eta \approx 2.246/2.3 \approx 0.977$.

Example 3.19

If $p_1 = 1$ and $p_i = 0$ for all $i > 1$, then $H_r(\mathcal{S}) = 0$. An r-ary optimal code $\mathcal{D}$ for $\mathcal{S}$ has average word-length $L(\mathcal{D}) = 1$, so here the upper bound $1 + H_r(\mathcal{S})$ is attained.

Exercise 3.4

Find the word-lengths, average word-length, and efficiency of a binary Shannon–Fano code for a source $\mathcal{S}$ with probabilities $p_i = 0.4, 0.3, 0.1, 0.1, 0.06, 0.04$. Compare this with Exercise 3.3, which concerns an optimal code for $\mathcal{S}$.

In general, Shannon–Fano codes are not far from optimal. They approach closer to optimality if we use them to encode extensions of sources (see §2.6). We will investigate the entropy of extensions in the next section, in preparation for a proof of this result in §3.6.

3.5 Entropy of Extensions and Products

Recall from §2.6 that S^n has q^n symbols $s_{i_1} \ldots s_{i_n}$, with probabilities $p_{i_1} \ldots p_{i_n}$. If we think of S^n as n independent copies of S then we should expect it to produce n times as much information as S. This suggests the following:

Theorem 3.20

If S is any source then $H_r(S^n) = nH_r(S)$.

Before proving this, we must first generalise the notion of an extension by considering products of sources. Let S and T be two sources, having symbols s_i and t_j with probabilities p_i and q_j; we define their *product* $S \times T$ to be the source whose symbols are the pairs (s_i, t_j), which we will abbreviate to $s_i t_j$, with probabilities $\Pr(s_i \text{ and } t_j)$. One can think of $S \times T$ as a pair consisting of S and T emitting symbols s_i and t_j simultaneously. We say that S and T are *independent* if $\Pr(s_i \text{ and } t_j) = p_i q_j$ for all i and j. For instance, S and T could represent the daily weather in two distant locations (but not nearby, since they would no longer be independent). The extension S^2 can be regarded as the product $S \times S$ of two independent copies of a single source S: a good example is a pair of identical but independent dice.

Lemma 3.21

If S and T are independent sources then $H_r(S \times T) = H_r(S) + H_r(T)$.

Proof

Independence gives $\Pr(s_i t_j) = p_i q_j$, so

$$
\begin{aligned}
H_r(S \times T) &= -\sum_i \sum_j p_i q_j \log_r p_i q_j \\
&= -\sum_i \sum_j p_i q_j (\log_r p_i + \log_r q_j) \\
&= -\sum_i \sum_j p_i q_j \log_r p_i - \sum_i \sum_j p_i q_j \log_r q_j \\
&= \left(-\sum_i p_i \log_r p_i \right) \left(\sum_j q_j \right) + \left(\sum_i p_i \right) \left(-\sum_j q_j \log_r q_j \right) \\
&= H_r(S) + H_r(T)
\end{aligned}
$$

since $\sum p_i = \sum q_j = 1$. $\qquad\square$

We can use induction to extend the definition of a product to any finite number of sources: we define

$$\mathcal{S}_1 \times \cdots \times \mathcal{S}_n = (\mathcal{S}_1 \times \cdots \times \mathcal{S}_{n-1}) \times \mathcal{S}_n.$$

The sources $\mathcal{S}_i$ are independent if each symbol $s_{i_1} \ldots s_{i_n}$ has probability $p_{i_1} \ldots p_{i_n}$, where each s_{i_j} has probability p_{i_j}.

Corollary 3.22

If $\mathcal{S}_1, \ldots, \mathcal{S}_n$ are independent sources then

$$H_r(\mathcal{S}_1 \times \cdots \times \mathcal{S}_n) = H_r(\mathcal{S}_1) + \cdots + H_r(\mathcal{S}_n).$$

Proof

This is proved by induction on n, using Lemma 3.21 for the inductive step. □

If we take $\mathcal{S}_1, \ldots, \mathcal{S}_n$ to be independent copies of $\mathcal{S}$, then $\mathcal{S}_1 \times \cdots \times \mathcal{S}_n = \mathcal{S}^n$, and Theorem 3.20 follows immediately from Corollary 3.22.

3.6 Shannon's First Theorem

Theorem 3.11 states that every uniquely decodable r-ary code $\mathcal{C}$ for a source S has average word-length $L(\mathcal{C}) \geq H_r(S)$, and Corollary 3.12 implies that this lower bound is not normally attained. However, we will show that the idea introduced at the end of §2.6, of using an optimal code for $\mathcal{S}^n$ as an encoding of $\mathcal{S}$, allows us to encode S with average word-lengths arbitrarily close to $H_r(\mathcal{S})$ as $n \to \infty$.

Recall that if a code for $\mathcal{S}^n$ has average word-length L_n, then as an encoding of S it has average word-length L_n/n. By Corollary 3.17, an optimal r-ary code for $\mathcal{S}^n$ has average word-length L_n satisfying

$$H_r(\mathcal{S}^n) \leq L_n \leq 1 + H_r(\mathcal{S}^n),$$

so Theorem 3.20 gives

$$nH_r(\mathcal{S}) \leq L_n \leq 1 + nH_r(\mathcal{S}).$$

Dividing by n we get

$$H_r(\mathcal{S}) \leq \frac{L_n}{n} \leq \frac{1}{n} + H_r(\mathcal{S}),$$

so

$$\lim_{n \to \infty} \frac{L_n}{n} = H_r(S).$$

This proves *Shannon's First Theorem*, or the *Noiseless Coding Theorem*, published by Shannon in his fundamental paper [Sh48]. Its full statement is:

Theorem 3.23

By encoding S^n with n sufficiently large, one can find uniquely decodable r-ary encodings of a source S with average word-lengths arbitrarily close to the entropy $H_r(S)$.

We considered a simple example of this in §2.6, for $n = 1, 2, 3$; we will return to this example, for arbitrary n, in the next section.

The "cost" of using this theorem is that, since $L_n/n \to H_r(S)$ rather slowly in many cases, we may need quite a large value of n to achieve efficient coding. Now if S has q symbols then S^n has q^n, a number which grows rapidly as n increases. This means that the construction of the code and the encoding process are both complicated and time-consuming. Also the decoding process involves delays while we wait for complete blocks of n symbols to be received, so we may have to compromise with a smaller value of n.

3.7 An Example of Shannon's First Theorem

Let S be a source with two symbols s_1, s_2 of probabilities $p_i = \frac{2}{3}, \frac{1}{3}$, as in Example 3.2. We saw in §3.1 that $H_2(S) = \log_2 3 - \frac{2}{3} \approx 0.918$, and in §2.6 we obtained the average word-lengths $L_n/n \approx 1$, 0.944 and 0.938 by using binary Huffman codes for S^n with $n = 1, 2$ and 3. For larger n it is simpler to use Shannon–Fano codes, rather than Huffman codes; they are a little less efficient, but they are easier to deal with, and they also give $L_n/n \to H_r(S)$ as $n \to \infty$.

There are 2^n symbols s for S^n, each consisting of a block of n symbols $s_i = s_1$ or s_2; if there are k symbols s_1 in s (and hence $n - k$ symbols s_2) then s has probability

$$\Pr(s) = \left(\frac{2}{3}\right)^k \left(\frac{1}{3}\right)^{n-k} = \frac{2^k}{3^n}.$$

For each $k = 0, 1, \ldots n$, the number of such symbols s is $\binom{n}{k}$, the number of ways of choosing k of the n symbols to be s_1. In Shannon–Fano coding (see §3.4), we assign to each such symbol s a code-word of length

$$l_k = \left\lceil \log_2 \left(\frac{1}{\Pr(s)}\right) \right\rceil = \left\lceil \log_2 \left(\frac{3^n}{2^k}\right) \right\rceil = \lceil n \log_2 3 - k \rceil = a_n - k,$$

where a_n denotes $\lceil n \log_2 3 \rceil$. Hence the average word-length (for encoding $\mathcal{S}^n$) is

$$L_n = \sum_{k=0}^{n} \binom{n}{k} \Pr(s) l_k$$

$$= \sum_{k=0}^{n} \binom{n}{k} \frac{2^k}{3^n} (a_n - k)$$

$$= \frac{1}{3^n} \left(a_n \sum_{k=0}^{n} \binom{n}{k} 2^k - \sum_{k=0}^{n} k \binom{n}{k} 2^k \right). \tag{3.9}$$

We can use the Binomial Theorem to evaluate the two summations in (3.9). We have

$$(1+x)^n = \sum_{k=0}^{n} \binom{n}{k} x^k, \tag{3.10}$$

so putting $x = 2$ we get

$$\sum_{k=0}^{n} \binom{n}{k} 2^k = 3^n.$$

(Alternatively, $\sum_k \binom{n}{k} 2^k / 3^n$ is the sum of the probabilities of the symbols s in $\mathcal{S}^n$, and hence equal to 1.) Differentiating (3.10) and then multiplying by x we have

$$nx(1+x)^{n-1} = \sum_{k=1}^{n} k \binom{n}{k} x^k = \sum_{k=0}^{n} k \binom{n}{k} x^k,$$

so putting $x = 2$ again we get

$$\sum_{k=0}^{n} k \binom{n}{k} 2^k = 2n.3^{n-1}.$$

Substituting in (3.9), we have

$$L_n = \frac{1}{3^n} \left(a_n 3^n - 2n.3^{n-1} \right) = a_n - \frac{2n}{3},$$

so

$$\frac{L_n}{n} = \frac{a_n}{n} - \frac{2}{3} = \frac{\lceil n \log_2 3 \rceil}{n} - \frac{2}{3}.$$

Now $n \log_2 3 \leq \lceil n \log_2 3 \rceil < 1 + n \log_2 3$, so

$$\log_2 3 \leq \frac{\lceil n \log_2 3 \rceil}{n} < \frac{1}{n} + \log_2 3,$$

giving

$$\frac{\lceil n \log_2 3 \rceil}{n} \to \log_2 3$$

and hence

$$\frac{L_n}{n} \to \log_2 3 - \frac{2}{3}$$

as $n \to \infty$. This limit is equal to $H_2(S) \approx 0.918$, so we have confirmed Shannon's Theorem for this particular source.

For $n = 1, \ldots, 10$, the average word-length $L = L_n/n$ is given in the following table, together with the efficiency $\eta = H/L$. (We use the approximation $\log_2 3 \approx 1.585$ to compute $a_n = \lceil n \log_2 3 \rceil$.)

n	1	2	3	4	5	6	7	8	9	10
a_n	2	4	5	7	8	10	12	13	15	16
L	1.333	1.333	1	1.083	0.933	1	1.048	0.958	1	0.933
η	0.689	0.689	0.918	0.848	0.984	0.918	0.876	0.959	0.918	0.984

This shows how $\eta \to 1$ (that is, $L \to H$) as $n \to \infty$, though convergence is rather slow and irregular. If, instead of Shannon–Fano codes, we use Huffman codes for S^n, we obtain the following table (restricted to $n \le 5$):

n	1	2	3	4	5
L	1	0.944	0.938	0.938	0.923
η	0.918	0.972	0.979	0.979	0.995

In this case, $\eta \to 1$ rather faster, though for certain values of n, such as $n = 5$, Shannon–Fano coding is almost as efficient as Huffman coding. When $n = 5$ this is because $3^n = 243 \approx 256 = 2^8$; thus the reciprocals of probabilities of the symbols in S^5 are only slightly less than powers of 2, so rounding up their logarithms with the ceiling function has only a small effect.

Exercise 3.5

Find the binary entropy $H_2(S)$, where S has two symbols with probabilities $\frac{4}{5}, \frac{1}{5}$. Find the average word-length L_n of a binary Shannon–Fano code for S^n, and verify that $\frac{1}{n}L_n \to H_2(S)$ as $n \to \infty$.

Exercise 3.6

Let S have q equiprobable symbols. Find the average word-length L_n of an r-ary Shannon–Fano code for S^n, and verify that $\frac{1}{n}L_n \to H_r(S)$ as $n \to \infty$.

3.8 Supplementary Exercises

Exercise 3.7

Let f be a strictly decreasing function $(0, 1] \to \mathbf{R}$ such that $f(ab) = f(a) + f(b)$ for all $a, b \in (0, 1]$. Show that $f(x) = -\log_r x$ for some $r > 1$, thus justifying the definition of the function I in (3.1). (Hint: consider the function $g(x) = f(e^{-x})$ for $x \geq 0$.)

Exercise 3.8

A source S consists of the sum of the scores of two independent unbiased dice. Find the probability distribution and the binary entropy of S, together with the average word-lengths of binary Huffman and Shannon–Fano codes for S.

Exercise 3.9

Draw the graphs of the functions $-p \log p$, $-\overline{p} \log \overline{p}$ and $H(p) = -p \log p - \overline{p} \log \overline{p}$ (which is the entropy of a source S with probabilities p and $\overline{p}$), for $0 \leq p \leq 1$, where $\log = \log_2$. Draw the graphs of the functions $p\lceil - \log p\rceil, \overline{p}\lceil - \log \overline{p}\rceil$ and $p\lceil - \log p\rceil + \overline{p}\lceil - \log \overline{p}\rceil$ (which is the average word-length $L(C)$ of a binary Shannon–Fano code C for S). Draw the graphs of $H(p), L(C)$ and $H(p) + 1$ on the same diagram, and check that this source satisfies Theorem 3.16.

Exercise 3.10

Show that if $q \geq 2$ then there is a source S with q symbols, and an instantaneous r-ary code C satisfying $L(C) = H_r(S)$, if and only if $q \equiv 1$ mod $(r - 1)$.

Exercise 3.11

Find the ternary entropy $H_3(S)$, where S has two symbols with probabilities $\frac{3}{4}, \frac{1}{4}$. Find the average word-length L_n of a ternary Shannon–Fano code for S^n, and verify that $\frac{1}{n}L_n \to H_3(S)$ as $n \to \infty$. Would a similar calculation work for *binary* Shannon–Fano codes for S^n ?

Exercise 3.12

Show that if $q \geq 2, r \geq 2$ and $\varepsilon > 0$ then there is a source S, with all probabilities $p_i > 0$, for which every instantaneous code C satisfies $L(C) > 1 + H_r(S) - \varepsilon$.

Exercise 3.13

How would you define the r-ary entropy $H_r(S)$ of a source S having infinitely many symbols of probabilities p_k $(k = 1, 2, 3, \ldots)$? Calculate $H_2(S)$ where $p_k = 2^{-k}$, and find an instantaneous binary code for this source with average word-length equal to $H_2(S)$.

Exercise 3.14

A source $\mathcal{S}$, emitting a sequence $X_1, X_2, \ldots$ of symbols $s_i \in S$, is a *Markov source* with a 1-symbol memory, meaning that we are given constant conditional probabilities $p_{ij} = \Pr(X_{n+1} = s_j \mid X_n = s_i)$, independent of n. If we assume that each $p_{ij} > 0$ then it can be shown that, over a long period, the symbols s_i have constant frequencies $p_i > 0$. Explain the definition $H(\mathcal{S}) = -\sum_i \sum_j p_i p_{ij} \log p_{ij}$ of the entropy of $\mathcal{S}$. Prove that $H(\mathcal{S}) \le H(\mathcal{T})$, where $\mathcal{T}$ is a memoryless source with symbols s_i and probabilities p_i, and determine when equality is attained. What is the interpretation of this result? Find (p_i), $H(\mathcal{S})$ and $H(\mathcal{T})$ when the probabilities p_{ij} are the entries in the matrix

$$(p_{ij}) = \frac{1}{6} \begin{pmatrix} 3 & 2 & 1 \\ 1 & 4 & 1 \\ 1 & 2 & 3 \end{pmatrix}.$$

4

Information Channels

The equivocation of the fiend that lies like truth. (*Macbeth*)

In this chapter we consider a source sending messages through an unreliable (or noisy) channel to a receiver. The "noise" in the channel could represent mechanical or human errors, or interference from another source. A good example is a space-probe, with a weak power-supply, sending back a message which has to be extracted from many other stronger competing signals. Because of noise, the symbols received may not be the same as those sent. Our aim here is to measure how much information is transmitted, and how much is lost in this process, using several different variations of the entropy function, and then to relate this to the average word-length of the code used.

4.1 Notation and Definitions

We will take the input of an information channel Γ to be a source $\mathcal{A}$, with a finite alphabet A of symbols $a = a_1, \ldots, a_r$, having probabilities

$$p_i = \Pr(a = a_i).$$

As usual, we require that

$$0 \le p_i \le 1 \qquad \text{and} \qquad \sum_{i=1}^{r} p_i = 1.$$

55

Here, $\mathcal{A}$ could be a source $\mathcal{S}$, with $a_i = s_i$ (the source-symbols), or alternatively $\mathcal{A}$ could represent a source $\mathcal{S}$ together with a code $\mathcal{C}$ for $\mathcal{S}$, in which case the symbols a_i could represent code-symbols t_j or code-words w_i. To allow for all these interpretations, we have changed the notation to $\mathcal{A}, A$ and a_i.

We will assume that whenever a symbol $a_i \in A$ is sent into the channel Γ, some symbol emerges from Γ. The output of Γ will be regarded as a source $\mathcal{B}$, with a finite alphabet B of symbols $b = b_1, \ldots, b_s$, having probabilities

$$q_j = \Pr(b = b_j)$$

where

$$0 \le q_j \le 1 \qquad \text{and} \qquad \sum_{j=1}^{s} q_j = 1.$$

Fig. 4.1 illustrates this situation.

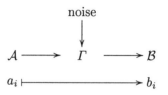

Figure 4.1

Example 4.1

In the *binary symmetric channel* (which we will abbreviate to BSC) we have $A = B = \mathbf{Z}_2 = \{0, 1\}$. Each input symbol $a = 0$ or 1 is correctly transmitted with probability P, and is incorrectly transmitted (as $\overline{a} = 1-a$) with probability $\overline{P} = 1 - P$, for some constant P $(0 \le P \le 1)$. This is illustrated in Fig. 4.2

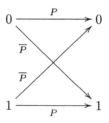

Figure 4.2

Example 4.2

In the *binary erasure channel* (BEC) we have $A = \mathbf{Z}_2 = \{0, 1\}$ and $B = \{0, 1, ?\}$. Each input symbol $a = 0$ or 1 is correctly transmitted with probability

P, and is erased (or made illegible) with probability $\overline{P}$, indicated by an output symbol $b = ?$ (see Fig. 4.3).

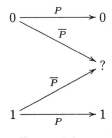

Figure 4.3

In general, we will assume that the behaviour of Γ is completely determined by its *forward probabilities*

$$P_{ij} = \Pr(b = b_j \mid a = a_i) = \Pr(b_j \mid a_i).$$

Thus P_{ij} is the conditional probability that the output symbol b is b_j, given that the corresponding input symbol is a_i. We assume that P_{ij} is independent of time, and of any previous symbols transmitted or received. If $a = a_i$, then b must be exactly one of the output symbols b_j, so

$$\sum_{j=1}^{s} P_{ij} = 1$$

for each $i = 1, \ldots, r$. These rs numbers P_{ij} form the *channel matrix*

$$M = (P_{ij}) = \begin{pmatrix} P_{11} & \cdots & P_{1s} \\ \vdots & & \vdots \\ P_{r1} & \cdots & P_{rs} \end{pmatrix},$$

which has r rows, indexed by the input symbols $a_1, \ldots, a_r$, and s columns, indexed by the output symbols $b_1, \ldots, b_s$; the entry in the i-th row and j-th column is P_{ij}. For instance, if Γ is the BSC or BEC we have

$$M = \begin{pmatrix} P & \overline{P} \\ \overline{P} & P \end{pmatrix} \quad \text{or} \quad \begin{pmatrix} P & 0 & \overline{P} \\ 0 & P & \overline{P} \end{pmatrix}.$$

The precise form of the channel matrix M depends on the ordering of the input symbols a_i and the output symbols b_j: a different ordering gives rise to a permutation of the rows or columns of M respectively. Thus the above matrix for the BEC uses the ordering $0, 1, ?$ of the output symbols, whereas the ordering $0, ?, 1$ would give

$$M = \begin{pmatrix} P & \overline{P} & 0 \\ 0 & \overline{P} & P \end{pmatrix}.$$

There are several ways of combining two channels Γ and Γ' to form a third channel. If Γ and Γ' have disjoint input alphabets A and A', and disjoint output alphabets B and B', then the *sum* $\Gamma + \Gamma'$ has input and output alphabets $A \cup A'$ and $B \cup B'$; each input symbol is transmitted through Γ or Γ' as it lies in A or A', so the channel matrix is a block matrix

$$\begin{pmatrix} M & O \\ O & M' \end{pmatrix}$$

where M and M' are the channel matrices for Γ and Γ'. There is an obvious extension to the sum of any finite number of channels.

In the case of the *product* $\Gamma \times \Gamma'$, we do not need to assume that A and A' or B and B' are disjoint. The input and output alphabets are $A \times A'$ and $B \times B'$, and the sender transmits a pair $(a, a') \in A \times A'$ by simultaneously sending a through Γ and a' through Γ', so that a pair $(b, b') \in B \times B'$ is received. Thus the forward probabilities are

$$\Pr\left((b, b') \mid (a, a')\right) = \Pr\left(b \mid a\right) . \Pr\left(b' \mid a'\right),$$

so the channel matrix is the *Kronecker product* $M \otimes M'$ of the matrices M and M' for Γ and Γ': if $M = (P_{ij})$ and $M' = (P'_{kl})$ are $r \times s$ and $r' \times s'$ matrices, then $M \otimes M'$ is an $rr' \times ss'$ matrix, with entries $P_{ij} P'_{kl}$. (The ordering of these entries depends on a choice of orderings for $A \times A'$ and $B \times B'$.) Again, one can extend this definition to the product of any finite number of channels; in particular, the *n-th extension* Γ^n of a channel Γ is the product $\Gamma \times \cdots \times \Gamma$ of n copies of Γ.

Example 4.3

If Γ and Γ' are binary symmetric channels, with channel matrices

$$M = \begin{pmatrix} P & \overline{P} \\ \overline{P} & P \end{pmatrix} \quad \text{and} \quad M' = \begin{pmatrix} P' & \overline{P'} \\ \overline{P'} & P' \end{pmatrix},$$

then $\Gamma + \Gamma'$ and $\Gamma \times \Gamma'$ have channel matrices

$$\begin{pmatrix} P & \overline{P} & 0 & 0 \\ \overline{P} & P & 0 & 0 \\ 0 & 0 & P' & \overline{P'} \\ 0 & 0 & \overline{P'} & P' \end{pmatrix} \quad \text{and} \quad \begin{pmatrix} PP' & \overline{P}P' & P\overline{P'} & \overline{P}\,\overline{P'} \\ \overline{P}P' & PP' & \overline{P}\,\overline{P'} & P\overline{P'} \\ P\overline{P'} & \overline{P}\,\overline{P'} & PP' & \overline{P}P' \\ \overline{P}\,\overline{P'} & P\overline{P'} & \overline{P}P' & PP' \end{pmatrix}.$$

(For $\Gamma \times \Gamma'$ we have used the ordering $(0,0), (1,0), (0,1), (1,1)$ of $A \times A' = B \times B' = \mathbf{Z}_2^2$.)

Exercise 4.1

The output of a channel Γ is used as the input for a channel Γ'. Find the channel matrix for the resulting composite channel $\Gamma \circ \Gamma'$, in terms of the matrices for Γ and Γ'. Generalise this result to the composition of any number of channels in series (this is called a *cascade* of channels).

Returning to the case of a single channel Γ, if we multiply the equations $\sum_i p_i = 1$ and $\sum_j P_{ij} = 1$ we get

$$\sum_{i=1}^{r}\sum_{j=1}^{s} p_i P_{ij} = 1 . \tag{4.1}$$

The probability that a_i is sent and b_j received is $p_i P_{ij}$. If b_j is received, then exactly one of the symbols a_i must have been sent, so we have the *channel relationships*

$$\sum_{i=1}^{r} p_i P_{ij} = q_j \qquad \text{for} \qquad j = 1, \ldots, s. \tag{4.2}$$

If we regard (p_i) as a vector $\mathbf{p} \in \mathbf{R}^r$, and (q_j) as a vector $\mathbf{q} \in \mathbf{R}^s$, then (4.2) can be written in the form

$$\mathbf{p}M = \mathbf{q} . \tag{4.2'}$$

If we sum (4.2) over all j, then reverse the order of summation, and use the fact that $\sum_j q_j = 1$, we obtain (4.1).

In addition to the forward probabilities P_{ij}, it is useful to define the *backward* probabilities

$$Q_{ij} = \Pr\left(a = a_i \mid b = b_j\right) = \Pr\left(a_i \mid b_j\right)$$

and the *joint probabilities*

$$R_{ij} = \Pr\left(a = a_i \text{ and } b = b_j\right) = \Pr\left(a_i, b_j\right) .$$

One can regard the forward probabilities P_{ij} as representing the point of view of the sender, who knows the input symbols a_i and is trying to guess the resulting output symbols b_j. Similarly, the backward probabilities Q_{ij} represent the point of view of the receiver, who knows the output symbols b_j and is trying to guess the corresponding input symbols a_i, while the joint probabilities R_{ij} represent an outside observer, who is trying to guess both a_i and b_j.

For every i and j we have

$$p_i P_{ij} = \Pr(a_i)\Pr\left(b_j \mid a_i\right) = \Pr\left(a_i, b_j\right) = \Pr\left(b_j\right)\Pr\left(a_i \mid b_j\right) = q_j Q_{ij} ,$$

all equal to R_{ij}, giving *Bayes' Formula*

$$Q_{ij} = \frac{p_i}{q_j} P_{ij} \tag{4.3}$$

provided $q_j \neq 0$. Combining this with (4.2) we get

$$Q_{ij} = \frac{p_i P_{ij}}{\sum_{k=1}^{r} p_k P_{kj}}. \tag{4.4}$$

We will consider some specific examples of these equations in the next section.

4.2 The Binary Symmetric Channel

One of the simplest and most frequently used information channels Γ is the binary symmetric channel (BSC), introduced in Example 4.1. In view of its importance, we will study it in more detail here. Recall that this channel is defined by:

(1) $A = B = \mathbf{Z}_2 = \{0, 1\}$,

(2) the channel matrix has the form

$$M = \begin{pmatrix} P_{00} & P_{01} \\ P_{10} & P_{11} \end{pmatrix} = \begin{pmatrix} P & \overline{P} \\ \overline{P} & P \end{pmatrix}$$

for some P where $0 \leq P \leq 1$. (For notational convenience, we use the subscripts $i, j = 0, 1$ rather than $1, 2$ here, so that $a_i = i$ and $b_j = j$ in the notation of §4.1.)

Condition (1) states that Γ is binary, and condition (2) states that Γ is symmetric (with respect to the symbols 0 and 1), in the sense that each input symbol a is correctly or incorrectly transmitted with probability P or $\overline{P}$, irrespective of whether $a = 0$ or 1.

The input probabilities have the form

$$p_0 = \Pr(a = 0) = p,$$
$$p_1 = \Pr(a = 1) = \overline{p},$$

for some p such that $0 \leq p \leq 1$. The channel relationships (4.2) then become

$$q_0 = \Pr(b = 0) = pP + \overline{p}\,\overline{P},$$
$$q_1 = \Pr(b = 1) = p\overline{P} + \overline{p}P;$$

writing $q_0 = q$ and $q_1 = \overline{q}$ we then have

$$(q, \overline{q}) = (p, \overline{p}) \begin{pmatrix} P & \overline{P} \\ \overline{P} & P \end{pmatrix},$$

as in equation (4.2'). If we substitute these values of q_j in Bayes' formula (4.3) we get:

$$Q_{00} = \frac{pP}{pP + \overline{p}\overline{P}}, \quad Q_{10} = \frac{\overline{p}\overline{P}}{pP + \overline{p}\overline{P}}, \quad Q_{01} = \frac{p\overline{P}}{p\overline{P} + \overline{p}P}, \quad Q_{11} = \frac{\overline{p}P}{p\overline{P} + \overline{p}P}.$$

Example 4.4

Let the input $\mathcal{A}$ be defined by putting $p = \frac{1}{2}$. Then $\overline{p} = \frac{1}{2}$, so the input symbols $a = 0$ and 1 are equiprobable. We have $q = \frac{1}{2}P + \frac{1}{2}\overline{P} = \frac{1}{2}(P + \overline{P}) = \frac{1}{2}$ and similarly $\overline{q} = \frac{1}{2}$, so the output symbols $b = 0$ and 1 are also equiprobable. The backward probabilities are given by

$$Q_{00} = Q_{11} = \frac{\frac{1}{2}P}{\frac{1}{2}} = P, \quad Q_{01} = Q_{10} = \frac{\frac{1}{2}\overline{P}}{\frac{1}{2}} = \overline{P}.$$

Example 4.5

Suppose that $P = 0.8$ (so Γ is fairly reliable, with 8 symbols out of 10 transmitted correctly), and $p = 0.9$ (so the input symbol a is almost always 0). Then we find that

$$q_0 = q = pP + \overline{p}\overline{P} = 0.74, \quad q_1 = \overline{q} = p\overline{P} + \overline{p}P = 0.26.$$

Thus the ouput symbol b is usually 0, but the bias towards 0 is not as strong as in the input. This loss of bias is due to noise, or errors, in the channel: these will cause an input $a = 0$ to be received as $b = 1$ more frequently than they cause 1 to be received as 0, simply because more symbols 0 are transmitted, so the proportion of 0s is reduced. The backward probabilities are

$$Q_{00} = \frac{p_0 P_{00}}{q_0} = \frac{0.9 \times 0.8}{0.74} \approx 0.973, \quad Q_{10} = \frac{p_1 P_{10}}{q_0} = \frac{0.1 \times 0.2}{0.74} \approx 0.027$$

(so if $b = 0$ then almost invariably $a = 0$),

$$Q_{01} = \frac{p_0 P_{01}}{q_1} = \frac{0.9 \times 0.2}{0.26} \approx 0.692, \quad Q_{11} = \frac{p_1 P_{11}}{q_1} = \frac{0.1 \times 0.8}{0.26} \approx 0.308$$

(so if $b = 1$ then usually $a = 0$!). Thus, no matter what symbol is received, the most likely input symbol was 0. The BSC behaves like this whenever both $Q_{00} > Q_{10}$ and $Q_{01} > Q_{11}$, that is, $pP > \overline{p}P = (1-p)\overline{P}$ and $p\overline{P} > \overline{p}P = (1-p)P$; we can write these two inequalities as $p = p(P+\overline{P}) > \overline{P}$ and $p = p(\overline{P}+P) > P$, or equivalently $p > \max(P, \overline{P})$. Similarly, if $\overline{p} > \max(P, \overline{P})$ then the most likely input symbol was 1.

Exercise 4.2

Let Γ be the BSC. Find necessary and sufficient conditions (on p and P) for Γ to satisfy

(i) $Q_{00} < Q_{10}$ and $Q_{01} < Q_{11}$;

(ii) $Q_{00} > Q_{10}$ and $Q_{01} < Q_{11}$;

(iii) $Q_{00} < Q_{10}$ and $Q_{01} > Q_{11}$.

What do these conditions mean, from the point of view of the receiver?

4.3 System Entropies

The input $\mathcal{A}$ and the output $\mathcal{B}$ of a channel Γ are sources with their own entropies; these are the *input entropy*

$$H(\mathcal{A}) = \sum_i p_i \log \frac{1}{p_i}$$

and the *output entropy*

$$H(\mathcal{B}) = \sum_j q_j \log \frac{1}{q_j}.$$

These represent the average amounts of information going into and coming out of Γ, per symbol, or equivalently, our uncertainty about the input and output.

Given that $b = b_j$ is received, there is a conditional entropy

$$H(\mathcal{A} \mid b_j) = \sum_i \Pr(a_i \mid b_j) \log \frac{1}{\Pr(a_i \mid b_j)} = \sum_i Q_{ij} \log \frac{1}{Q_{ij}};$$

this represents the receiver's uncertainty about $\mathcal{A}$, given that b_j is received, or equivalently, how much more information he would gain by knowing $\mathcal{A}$. Averaging over all b_j, and using $q_j Q_{ij} = R_{ij}$, we get the *equivocation* (of $\mathcal{A}$ with respect to $\mathcal{B}$)

$$H(\mathcal{A} \mid \mathcal{B}) = \sum_j q_j H(\mathcal{A} \mid b_j) = \sum_j q_j \left(\sum_i Q_{ij} \log \frac{1}{Q_{ij}} \right) = \sum_i \sum_j R_{ij} \log \frac{1}{Q_{ij}}.$$

This represents the receiver's average uncertainty about $\mathcal{A}$ when receiving $\mathcal{B}$, or equivalently, how much extra information would be gained by also knowing $\mathcal{A}$. Similarly, if a_i is sent then the uncertainty about $\mathcal{B}$ is the conditional entropy

$$H(\mathcal{B} \mid a_i) = \sum_j \Pr(b_j \mid a_i) \log \frac{1}{\Pr(b_j \mid a_i)} = \sum_j P_{ij} \log \frac{1}{P_{ij}}.$$

Averaging over all a_i, and using $p_i P_{ij} = R_{ij}$, we get

$$H(\mathcal{B} \mid \mathcal{A}) = \sum_i p_i H(\mathcal{B} \mid a_i) = \sum_i p_i \left(\sum_j P_{ij} \log \frac{1}{P_{ij}} \right) = \sum_i \sum_j R_{ij} \log \frac{1}{P_{ij}} ;$$

this is the equivocation of $\mathcal{B}$ with respect to $\mathcal{A}$, representing the sender's average uncertainty about $\mathcal{B}$ when $\mathcal{A}$ is known, or equivalently, how much extra information would be gained by also knowing $\mathcal{B}$.

An observer trying to guess both the input and the output of Γ will have average uncertainty given by the *joint entropy*

$$H(\mathcal{A}, \mathcal{B}) = \sum_i \sum_j \Pr\left(a_i, b_j\right) \log \frac{1}{\Pr\left(a_i, b_j\right)} = \sum_i \sum_j R_{ij} \log \frac{1}{R_{ij}} .$$

If Γ is such that $\mathcal{A}$ and $\mathcal{B}$ are statistically independent, that is, if $R_{ij} = p_i q_j$ for all i and j (unlikely in real life!), then we have

$$H(\mathcal{A}, \mathcal{B}) = \sum_i \sum_j p_i q_j \left(\log \frac{1}{p_i} + \log \frac{1}{q_j} \right)$$

$$= \sum_i p_i \log \frac{1}{p_i} + \sum_j q_j \log \frac{1}{q_j} \qquad \left(\text{since} \quad \sum_i p_i = \sum_j q_j = 1 \right)$$

$$= H(\mathcal{A}) + H(\mathcal{B}) . \tag{4.5}$$

Thus, in this case, the information conveyed by $\mathcal{A}$ and $\mathcal{B}$ together is the sum of the amounts they convey separately (in other cases, we shall see that it is less). If we think of entropy as measuring an amount of information (or uncertainty), then (4.5) is analogous to the result that $|A \cup B| = |A| + |B|$ for disjoint finite sets A and B.

In general, one would expect $\mathcal{A}$ and $\mathcal{B}$ to be related, rather than independent, so in such cases we use $R_{ij} = p_i P_{ij}$ to give

$$H(\mathcal{A}, \mathcal{B}) = \sum_i \sum_j R_{ij} \log \frac{1}{p_i} + \sum_i \sum_j R_{ij} \log \frac{1}{P_{ij}} .$$

Now $\sum_j R_{ij} = p_i$ for each i, so

$$H(\mathcal{A}, \mathcal{B}) = \sum_i p_i \log \frac{1}{p_i} + \sum_i \sum_j R_{ij} \log \frac{1}{P_{ij}}$$

$$= H(\mathcal{A}) + H(\mathcal{B} \mid \mathcal{A}) . \tag{4.6}$$

This confirms the interpretation of $H(\mathcal{B} \mid \mathcal{A})$ as the extra information conveyed by $\mathcal{B}$ when $\mathcal{A}$ is already known. It is analogous to the rule $|A \cup B| = |A| + |B \setminus A|$ for finite sets A and B. By a similar argument, transposing the roles of $\mathcal{A}$ and $\mathcal{B}$, we have

$$H(\mathcal{A}, \mathcal{B}) = H(\mathcal{B}) + H(\mathcal{A} \mid \mathcal{B}) , \tag{4.7}$$

with a similar interpretation of $H(\mathcal{A} \mid \mathcal{B})$; this corresponds to $|A \cup B| = |B| + |A \setminus B|$.

We call $H(\mathcal{A})$, $H(\mathcal{B})$, $H(\mathcal{A} \mid \mathcal{B})$, $H(\mathcal{B} \mid \mathcal{A})$ and $H(\mathcal{A}, \mathcal{B})$ the *system entropies*; they depend on both Γ and $\mathcal{A}$ (which, between them, determine $\mathcal{B}$).

Exercise 4.3

Prove equation (4.7), that $H(\mathcal{A}, \mathcal{B}) = H(\mathcal{B}) + H(\mathcal{A} \mid \mathcal{B})$. What interpretation of the equivocation $H(\mathcal{A} \mid \mathcal{B})$ does this imply?

Exercise 4.4

Show that the system entropies of a product channel $\Gamma \times \Gamma'$ are obtained by adding those for Γ and Γ', while the system entropies for the n-th extension Γ^n are n times those for Γ. (Hint: see §3.5.)

4.4 System Entropies for the Binary Symmetric Channel

Let Γ be the BSC, with the notation as in §4.2. The input and output entropies are

$$H(\mathcal{A}) = -p \log p - \bar{p} \log \bar{p} = H(p),$$
$$H(\mathcal{B}) = -q \log q - \bar{q} \log \bar{q} = H(q),$$

where $q = pP + \bar{p}\bar{P}$. To compare these, we use convexity.

A function $f : [0, 1] \to \mathbf{R}$ is *strictly convex* if, whenever $a, b \in [0, 1]$ and $x = \lambda a + \bar{\lambda} b$ with $0 \le \lambda \le 1$, then

$$f(x) \ge \lambda f(a) + \bar{\lambda} f(b),$$

with equality if and only if $x = a$ or b, that is, $a = b$ or $\lambda = 0$ or 1. Since $\bar{\lambda} = 1 - \lambda$, x ranges from b to a as λ varies between 0 and 1; the graph of $\lambda f(a) + \bar{\lambda} f(b)$ is the straight line joining the points $(a, f(a))$ and $(b, f(b))$, so convexity means that, between any points a and b in the domain, the graph of f is above this straight line, as shown in Fig. 4.4.[1]

The graph of the function $H(p)$, shown in Fig. 3.3, suggests that this function is strictly convex. We need to prove this, in order to deduce some important inequalities involving entropy. First we need a general result from Calculus.

---

[1] In some areas of mathematics, such as Analysis and Operations Research, the main inequality in this definition is reversed, so the graph is *below* the line.

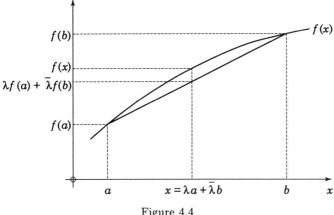

Figure 4.4

Lemma 4.6

If a function $f : [0,1] \rightarrow \mathbf{R}$ is continuous on the interval $[0,1]$ and twice differentiable on $(0,1)$, with $f''(x) < 0$ for all $x \in (0,1)$, then f is strictly convex.

Exercise 4.5

Prove Lemma 4.6, using the Mean Value Theorem. (Hint: this states that if a function g is continuous on $[0,1]$ and differentiable on $(0,1)$, and if $0 \leq a < b \leq 1$, then $(g(b) - g(a))/(b - a) = g'(c)$ for some c between a and b; see [Fi83] or [La83], for instance. Assume that the lemma is false, and obtain a contradiction by applying the Mean Value Theorem three times: twice to f and then once to f'.)

Corollary 4.7

The entropy function $H(p)$ is strictly convex on $[0,1]$.

Proof

We have $H(p) = -p \log p - (1-p) \log(1-p)$ for $0 < p < 1$, so $H(p)$ is continuous and twice differentiable on $(0,1)$; the convention that $H(0) = H(1) = 0$ means that it is continuous on $[0,1]$. Without loss of generality we can assume that the logarithms are natural logarithms, so $H'(p) = -\ln p + \ln(1-p)$ and hence

$$H''(p) = -\frac{1}{p} - \frac{1}{1-p} < 0$$

for all $p \in (0,1)$. The result now follows from Lemma 4.6. $\qquad \square$

See Exercise 4.14 for an extension of this result to sources with an arbitrary number of symbols.

We now return to the BSC. In the definition of strict convexity, if we take $a = p$, $b = \overline{p}$ and $\lambda = P$, we see that $x = pP + \overline{p}\overline{P} = q$, so $H(q) \geq H(p)$ with equality if and only if $p = \overline{p}$ (that is, $p = \frac{1}{2}$), or $q = \overline{p}$ or p (that is, $P = 0$ or 1). Since $H(\mathcal{A}) = H(p)$ and $H(\mathcal{B}) = H(q)$, this implies that the BSC satisfies

$$H(\mathcal{B}) \geq H(\mathcal{A}), \tag{4.8}$$

with equality if and only if the input symbols are equiprobable ($p = \frac{1}{2}$) or the channel is totally unreliable ($P = 0$) or reliable ($P = 1$). The inequality (4.8), illustrated in Fig. 4.5, shows that transmission through the BSC generally increases uncertainty; however, there are some channels for which this is not true (see Exercise 4.6).

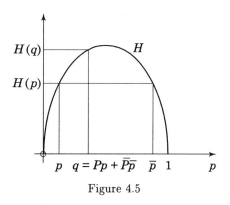

Figure 4.5

Exercise 4.6

Give an example of an information channel Γ and an input $\mathcal{A}$ for which $H(\mathcal{B}) < H(\mathcal{A})$, where $\mathcal{B}$ is the resulting output.

For the BSC we have

$$\begin{aligned}
H(\mathcal{B} \mid \mathcal{A}) &= \sum_i \sum_j p_i P_{ij} \log \frac{1}{P_{ij}} \\
&= -pP \log P - p\overline{P} \log \overline{P} - \overline{p}\overline{P} \log \overline{P} - \overline{p}P \log P \\
&= -(p + \overline{p})P \log P - (p + \overline{p})\overline{P} \log \overline{P} \\
&= -P \log P - \overline{P} \log \overline{P} \\
&= H(P).
\end{aligned}$$

Thus the sender's uncertainty about the output is equal to the uncertainty as to whether symbols are transmitted correctly. This is least when $P = 0$ or 1 (Γ is totally unreliable or reliable), and greatest when $P = \frac{1}{2}$ (Γ is useless).

Equation (4.6) implies that

$$H(\mathcal{A}, \mathcal{B}) = H(\mathcal{A}) + H(\mathcal{B} \mid \mathcal{A}) = H(p) + H(P),$$

while Equation (4.7) implies that

$$H(\mathcal{A}, \mathcal{B}) = H(\mathcal{B}) + H(\mathcal{A} \mid \mathcal{B}) = H(q) + H(\mathcal{A} \mid \mathcal{B}),$$

so the equivocation for the BSC is given by

$$H(\mathcal{A} \mid \mathcal{B}) = H(p) + H(P) - H(q).$$

Now $q = pP + \bar{p}\bar{P}$, which lies between P and $\bar{P}$ since $0 \le p \le 1$, so $H(q) \ge H(P)$ by Corollary 4.7. It immediately follows that the BSC satisfies

$$H(\mathcal{B} \mid \mathcal{A}) \le H(\mathcal{B}), \tag{4.9}$$

with equality if and only if $P = \frac{1}{2}$ or $p = 0, 1$. This means that the uncertainty about $\mathcal{B}$ generally decreases when $\mathcal{A}$ is known. Similarly, $H(\mathcal{A} \mid \mathcal{B}) = H(p) + H(P) - H(q) \le H(p)$ (since $H(q) \ge H(P)$), so

$$H(\mathcal{A} \mid \mathcal{B}) \le H(\mathcal{A}), \tag{4.10}$$

with equality if and only if $P = \frac{1}{2}$ or $p = 0, 1$. This means that the uncertainty about $\mathcal{A}$ generally decreases when $\mathcal{B}$ is known. We shall see later that the inequalities (4.9) and (4.10) are valid for all channels, not just the BSC.

Exercise 4.7

Calculate the system entropies where Γ is the binary erasure channel (BEC), introduced in §4.1, and the input probabilities of 0 and 1 are p and $\bar{p}$. Show that this channel satisfies (4.9) and (4.10).

4.5 Extension of Shannon's First Theorem to Information Channels

Shannon's First Theorem (for sources, Theorem 3.23) states that the greatest lower bound of the average word-lengths of uniquely decodable encodings of a source $\mathcal{A}$ is equal to the entropy $H(\mathcal{A})$. Similarly, Shannon's First Theorem for channels (which we will prove in this section) states that the greatest lower bound of the average word-lengths of uniquely decodable encodings of the input $\mathcal{A}$ of a channel, given knowledge of its output $\mathcal{B}$, is equal to the equivocation $H(\mathcal{A} \mid \mathcal{B})$. In each of these two cases, least average word-length is an accurate measure of information. In the case of a channel Γ, we interpret this from the point of view of the receiver, who knows $\mathcal{B}$ but is uncertain about $\mathcal{A}$; the extra information needed to be certain about $\mathcal{A}$ is the equivocation $H(\mathcal{A} \mid \mathcal{B})$, and this is equal to the least average word-length required to supply that extra information (by some other means, separate from Γ). In effect, the receiver is saying "I know $\mathcal{B}$, but I am not sure about $\mathcal{A}$; give me more information, so that I know $\mathcal{A}$", and we are trying to measure the extra information required.

Suppose that $b = b_j$ is received; knowing this, how can this extra information about the input symbol $a = a_i$ best be encoded? As in Chapter 3, we will use Shannon–Fano coding of extensions $\mathcal{A}^n$ of $\mathcal{A}$, the only difference being that we now use the *conditional* probabilities $Q_{ij} = \Pr(a_i \mid b_j)$ for $\mathcal{A}$ rather than the unconditional probabilities $p_i = \Pr(a_i)$, since we now know b_j.

For simplicity, let us first take $n = 1$, so (knowing that $b = b_j$) we construct a Shannon–Fano code for $\mathcal{A}$ as in §3.4. This is an instantaneous r-ary code $\mathcal{C}_j$ for $\mathcal{A}$ with average word-length $L_{(j)}$ satisfying

$$H(\mathcal{A} \mid b_j) \le L_{(j)} \le 1 + H(\mathcal{A} \mid b_j) \tag{4.11}$$

by Theorem 3.16, where

$$H(\mathcal{A} \mid b_j) = \sum_i Q_{ij} \log \frac{1}{Q_{ij}}.$$

We now form an encoding $\mathcal{C}$ for $\mathcal{A}$, using the code $\mathcal{C}_j$ whenever b_j is received, so that $\mathcal{C}$ can be regarded as an "average" of the codes $\mathcal{C}_j$. Taking the average of the terms in (4.11) over all $b_j \in \mathcal{B}$ (with probabilities q_j), we see that $\mathcal{C}$ has average word-length $L = \sum_j q_j L_{(j)}$ satisfying

$$H(\mathcal{A} \mid \mathcal{B}) \le L \le 1 + H(\mathcal{A} \mid \mathcal{B}). \tag{4.12}$$

Now recall from §4.1 that the n-th extension Γ^n of Γ is a channel with input and output alphabets A^n and B^n. Each word $a_{i_1} \ldots a_{i_n} \in A^n$ is transmitted through Γ^n by sending its symbols $a_{i_1}, \ldots, a_{i_n}$ in succession through Γ, or equivalently by sending them simultaneously through n independent copies of Γ, so the forward probabilities of Γ^n have the form

$$\Pr(b_{i_1} \ldots b_{i_n} \mid a_{i_1} \ldots a_{i_n}) = \Pr(b_{i_1} \mid a_{i_1}) \ldots \Pr(b_{i_n} \mid a_{i_n}).$$

If we use $\mathcal{A}^n$ to define the input probability distribution for Γ^n, then the output distribution is given by $\mathcal{B}^n$. Theorem 3.20 gives

$$H(\mathcal{A}^n) = nH(\mathcal{A}) \quad \text{and} \quad H(\mathcal{B}^n) = nH(\mathcal{B}), \tag{4.13}$$

and similarly Exercise 4.4 gives

$$H(\mathcal{A}^n \mid \mathcal{B}^n) = nH(\mathcal{A} \mid \mathcal{B}),$$
$$H(\mathcal{B}^n \mid \mathcal{A}^n) = nH(\mathcal{B} \mid \mathcal{A}), \qquad (4.14)$$
$$H(\mathcal{A}^n, \mathcal{B}^n) = nH(\mathcal{A}, \mathcal{B}).$$

If we apply the idea of averaging Shannon–Fano codes to $\mathcal{A}^n$ rather than to $\mathcal{A}$, then by (4.12) we get an encoding of $\mathcal{A}^n$ with average word-length L_n satisfying

$$H(\mathcal{A}^n \mid \mathcal{B}^n) \le L_n \le 1 + H(\mathcal{A}^n \mid \mathcal{B}^n),$$

so that (4.14) gives

$$nH(\mathcal{A} \mid \mathcal{B}) \le L_n \le 1 + nH(\mathcal{A} \mid \mathcal{B}).$$

As an encoding of $\mathcal{A}$, this is uniquely decodable and has average word-length L_n/n. If we divide by n we have

$$H(\mathcal{A} \mid \mathcal{B}) \le \frac{L_n}{n} \le \frac{1}{n} + H(\mathcal{A} \mid \mathcal{B})$$

for all n, so

$$\frac{L_n}{n} \to H(\mathcal{A} \mid \mathcal{B}) \quad \text{as} \quad n \to \infty.$$

This proves Shannon's Theorem (the analogue of Theorem 3.23):

Theorem 4.8

If the output $\mathcal{B}$ of a channel is known, then by encoding $\mathcal{A}^n$ with n sufficiently large, one can find uniquely decodable encodings of the input $\mathcal{A}$ with average word-lengths arbitrarily close to the equivocation $H(\mathcal{A} \mid \mathcal{B})$.

As in the case of source coding, it can be shown that the average word-length can never be lower than this bound. Theorems 4.8 and 3.23 show that $H(\mathcal{A} \mid \mathcal{B})$ and $H(\mathcal{A})$ represent the information conveyed by $\mathcal{A}$, when $\mathcal{B}$ is respectively known or not known, as measured in average word-length. One would therefore expect every channel to satisfy $H(\mathcal{A} \mid \mathcal{B}) \le H(\mathcal{A})$, since one cannot learn more from $\mathcal{A}$ when $\mathcal{B}$ is known than when it is unknown. We proved this inequality for the BSC in statement (4.10), and in the next section we will prove it in general.

4.6 Mutual Information

If Γ is a channel with input $\mathcal{A}$ and output $\mathcal{B}$, then the entropy $H(\mathcal{A})$ of $\mathcal{A}$ has three equivalent interpretations:

(1) it is the uncertainty about $\mathcal{A}$ when $\mathcal{B}$ is unknown;

(2) it is the information conveyed by $\mathcal{A}$ when $\mathcal{B}$ is unknown;

(3) it is the average word-length needed to encode $\mathcal{A}$ when $\mathcal{B}$ is unknown.

Similarly, the equivocation $H(\mathcal{A} \mid \mathcal{B})$ has three equivalent interpretations:

(1) it is the uncertainty about $\mathcal{A}$ when $\mathcal{B}$ is known;

(2) it is the information conveyed by $\mathcal{A}$ when $\mathcal{B}$ is known;

(3) it is the average word-length needed to encode $\mathcal{A}$ when $\mathcal{B}$ is known.

We define the difference between these two numbers to be the *mutual information*

$$I(\mathcal{A}, \mathcal{B}) = H(\mathcal{A}) - H(\mathcal{A} \mid \mathcal{B}).$$

This also has three equivalent interpretations, analogous to those above:

(1) it is the amount of uncertainty about $\mathcal{A}$ resolved by knowing $\mathcal{B}$;

(2) it is the amount of information about $\mathcal{A}$ conveyed by $\mathcal{B}$;

(3) it is the average number of symbols, in the code-words for $\mathcal{A}$, which refer to $\mathcal{B}$.

In their different ways, these interpretations all show that $I(\mathcal{A}, \mathcal{B})$ represents how much information $\mathcal{A}$ and $\mathcal{B}$ have in common. If we continue the analogy with finite sets used in §4.3, we can think of $I(\mathcal{A}, \mathcal{B})$ as corresponding to the intersection of sets, since $|A \cap B| = |A| - |A \setminus B|$.

Example 4.9

For a rather frivolous example, let Γ be a film company, $\mathcal{A}$ a book, and $\mathcal{B}$ the resulting film of the book. Then $I(\mathcal{A}, \mathcal{B})$ represents how much the film tells you about the book.

Example 4.10

Let $\mathcal{A}$ be a lecture, Γ a student taking notes, and $\mathcal{B}$ the resulting set of lecture-notes. Then $I(\mathcal{A}, \mathcal{B})$ measures how accurately the notes record the lecture.

Interchanging the roles of $\mathcal{A}$ and $\mathcal{B}$, we can define

$$I(\mathcal{B}, \mathcal{A}) = H(\mathcal{B}) - H(\mathcal{B} \mid \mathcal{A}),$$

the amount of information about $\mathcal{B}$ conveyed by $\mathcal{A}$ (for instance, how much reading the book tells you about the film). This is analogous to $|B \cap A| = |B| - |B \setminus A|$. We saw in (4.6) and (4.7) that

$$H(\mathcal{A}, \mathcal{B}) = H(\mathcal{A}) + H(\mathcal{B} \mid \mathcal{A}),$$
$$H(\mathcal{A}, \mathcal{B}) = H(\mathcal{B}) + H(\mathcal{A} \mid \mathcal{B}).$$

Eliminating $H(\mathcal{A}, \mathcal{B})$ from these two equations, we see that

$$H(\mathcal{A}) - H(\mathcal{A} \mid \mathcal{B}) = H(\mathcal{B}) - H(\mathcal{B} \mid \mathcal{A}),$$

so

$$I(\mathcal{A}, \mathcal{B}) = I(\mathcal{B}, \mathcal{A}). \tag{4.15}$$

Thus the output tells you exactly as much about the input as the input tells you about the output. If we use (4.7) to substitute for $H(\mathcal{A} \mid \mathcal{B})$ in the definition of $I(\mathcal{A}, \mathcal{B})$, we get

$$I(\mathcal{A}, \mathcal{B}) = H(\mathcal{A}) + H(\mathcal{B}) - H(\mathcal{A}, \mathcal{B}). \tag{4.16}$$

(This is analogous to $|A \cap B| = |A| + |B| - |A \cup B|$, just as (4.15) corresponds to $|A \cap B| = |B \cap A|$.)

Theorem 4.11

For every channel Γ we have $I(\mathcal{A}, \mathcal{B}) \geq 0$, with equality if and only if the input $\mathcal{A}$ and the output $\mathcal{B}$ are statistically independent.

Proof

Equation (4.16) gives

$$
\begin{aligned}
I(\mathcal{A}, \mathcal{B}) &= H(\mathcal{A}) + H(\mathcal{B}) - H(\mathcal{A}, \mathcal{B}) \\
&= \sum_i p_i \log \frac{1}{p_i} + \sum_j q_j \log \frac{1}{q_j} - \sum_i \sum_j R_{ij} \log \frac{1}{R_{ij}} \\
&= \sum_i \sum_j R_{ij} \log \frac{1}{p_i} + \sum_i \sum_j R_{ij} \log \frac{1}{q_j} - \sum_i \sum_j R_{ij} \log \frac{1}{R_{ij}} \\
&= \sum_i \sum_j R_{ij} \log \frac{1}{p_i q_j} - \sum_i \sum_j R_{ij} \log \frac{1}{R_{ij}},
\end{aligned}
$$

where we have used the facts that $p_i = \sum_j R_{ij}$ and $q_j = \sum_i R_{ij}$. Now $\sum_i \sum_j R_{ij} = \sum_i \sum_j p_i q_j = 1$, so we can apply Corollary 3.9 to the probability distributions (R_{ij}) and $(p_i q_j)$. (We may assume that each $p_i q_j > 0$, as required in Corollary 3.9, by ignoring any input or output symbols with zero probability, without changing the system entropies.) Corollary 3.9 shows that

$$\sum_i \sum_j R_{ij} \log \frac{1}{p_i q_j} \geq \sum_i \sum_j R_{ij} \log \frac{1}{R_{ij}},$$

as required, with equality if and only if $R_{ij} = p_i q_j$ for all i and j, that is, if and only if $\mathcal{A}$ and $\mathcal{B}$ are statistically independent. $\qquad\square$

Corollary 4.12

For every channel Γ we have

$$H(\mathcal{A}) \geq H(\mathcal{A} \mid \mathcal{B}), \quad H(\mathcal{B}) \geq H(\mathcal{B} \mid \mathcal{A}) \quad \text{and} \quad H(\mathcal{A}, \mathcal{B}) \leq H(\mathcal{A}) + H(\mathcal{B});$$

in each case, there is equality if and only if the input $\mathcal{A}$ and the output $\mathcal{B}$ are statistically independent.

Proof

This follows immediately from Theorem 4.11, using the equations

$$\begin{aligned} I(\mathcal{A}, \mathcal{B}) &= H(\mathcal{A}) - H(\mathcal{A} \mid \mathcal{B}) \\ &= H(\mathcal{B}) - H(\mathcal{B} \mid \mathcal{A}) \\ &= H(\mathcal{A}) + H(\mathcal{B}) - H(\mathcal{A}, \mathcal{B}) \end{aligned}$$

proved earlier in this section. $\qquad\square$

We will give a simple illustration of these results in the next section.

4.7 Mutual Information for the Binary Symmetric Channel

As an example of calculating mutual information, let us take the channel Γ to be the BSC, with the usual notation (see §4.2). In §4.6 we saw that the mutual information of any channel is given by

$$I(\mathcal{A}, \mathcal{B}) = H(\mathcal{B}) - H(\mathcal{B} \mid \mathcal{A}).$$

In §4.4 we saw that the BSC has $H(\mathcal{B}) = H(q)$ and $H(\mathcal{B} \mid \mathcal{A}) = H(P)$, where $q = pP + \bar{p}\bar{P}$, so

$$\begin{aligned} I(\mathcal{A}, \mathcal{B}) &= H(q) - H(P) \\ &= H(pP + \bar{p}\bar{P}) - H(P). \end{aligned}$$

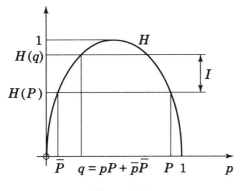

Figure 4.6

Note that this depends on both P and p, that is, on the channel and its input. A graph of the function H (see Fig. 4.6) shows that $0 \leq I(\mathcal{A}, \mathcal{B}) \leq 1 - H(P)$.

For a fixed P, $I(\mathcal{A}, \mathcal{B})$ attains its maximum value $1 - H(P)$ when $p = \frac{1}{2}$ (that is, $q = \frac{1}{2}$), and its minimum value 0 when $p = 0$ or 1. These extremes correspond to the cases where the input symbols are respectively equiprobable or constant.

Exercise 4.8

Let $I = I(\mathcal{A}, \mathcal{B})$ be the mutual information for the BSC; sketch the surface in $\mathbf{R}^3$ representing I as a function of p and P. Do the same for the BEC.

4.8 Channel Capacity

The mutual information $I(\mathcal{A}, \mathcal{B})$ for a channel Γ represents how much of the information in the input $\mathcal{A}$ is emerging in the output $\mathcal{B}$. This depends on both Γ and $\mathcal{A}$, as we saw in §4.7 in the case of the BSC. For a given channel Γ, we wish to maximise this by choosing the source $\mathcal{A}$ suitably (or by taking the input to be a suitable encoding of a given source).

We define the *capacity* C of a channel Γ to be the maximum value of the mutual information $I(\mathcal{A}, \mathcal{B})$, where $\mathcal{A}$ ranges over all possible inputs for Γ; thus we keep the forward probabilities P_{ij} fixed, while the input probabilities p_i are allowed to vary. This means that C, which depends on Γ alone, represents the maximum amount of information which the channel can transmit.

Example 4.13

We saw at the end of §4.7 that the BSC has channel capacity $C = 1 - H(P)$, attained when the input satisfies $p = \frac{1}{2}$. Figure 4.7 shows C as a function of P. Notice that C is greatest when P is 0 or 1, that is, Γ is completely reliable or unreliable (and thus completely predictable); C is least when $P = \frac{1}{2}$, that is, when Γ is most unpredictable.

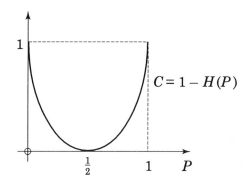

Figure 4.7

Exercise 4.9

Find the mutual information of the BEC, using Exercise 4.7, and hence find the capacity of this channel.

Exercise 4.10

Show that if channels Γ and Γ' have capacities C and C', then their product $\Gamma \times \Gamma'$ has capacity $C + C'$, and the extension Γ^n has capacity nC.

For a general channel Γ, calculating C can be difficult: this is because one is required to maximise a function $I(\mathcal{A}, \mathcal{B})$ which is non-linear, involving logarithms of probabilities, subject to the constraint $\sum p_i = 1$ (or equivalently $\sum q_j = 1$). See Exercise 4.16 for a simple example of this.

To justify our definition of capacity, we need to show that C always exists, that is, that for every channel Γ the mutual information $I(\mathcal{A}, \mathcal{B})$ is bounded above and attains its least upper bound. (It is conceivable that it could be unbounded, or bounded above without attaining its least upper bound.) To do this, we will use some ideas and results from Analysis; readers who are allergic to this topic, or who are prepared to take the existence of C for granted, can skip the rest of this section.

Let us keep Γ fixed and vary $\mathcal{A}$, so we keep the channel matrix $M = (P_{ij})$ constant and let the input probability distribution vector $\mathbf{p} = (p_1, \ldots, p_r)$ vary, ranging over the set

$$\mathcal{P} = \{\mathbf{p} \in \mathbf{R}^r \mid p_i \geq 0, \sum_i p_i = 1\}$$

of all probability distribution vectors with r components. Geometrically, $\mathcal{P}$ is an $(r-1)$-dimensional simplex, the convex set bounded by the r standard basis vectors in $\mathbf{R}^r$: when $r = 2$ or 3 it is respectively a line-segment or a triangle (see Fig. 4.8), and when $r = 4$ it is a tetrahedron in $\mathbf{R}^4$.

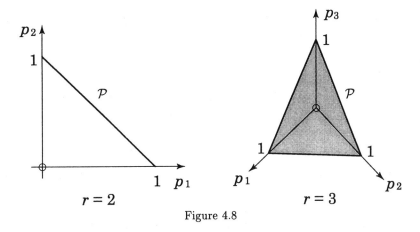

Figure 4.8

A subset $X \subseteq \mathbf{R}^r$ is *closed* if X contains the limit $\lim_{n \to \infty} \mathbf{x}_n$ of each convergent sequence of points $\mathbf{x}_n \in X$; equivalently, if $\mathbf{y} \notin X$ then every point sufficiently close to $\mathbf{y}$ is also outside X. We say that X is *bounded* if there is some real number M such that $|\mathbf{x}| \leq M$ for all $\mathbf{x} \in X$. A closed, bounded subset $X \subseteq \mathbf{R}^r$ is said to be *compact*.[2] It is straightforward (see Exercise 4.11) to show that $\mathcal{P}$ is compact for each r.

Exercise 4.11

Prove that the set $\mathcal{P}$ of probability distribution vectors $\mathbf{p} = (p_1, \ldots, p_r)$ is a closed and bounded subset of $\mathbf{R}^r$.

We saw in §4.6 that $I(\mathcal{A}, \mathcal{B}) = H(\mathcal{B}) - H(\mathcal{B} \mid \mathcal{A})$. Now

$$H(\mathcal{B} \mid \mathcal{A}) = \sum_i \sum_j p_i P_{ij} \log \frac{1}{P_{ij}} = \sum_i \left(\sum_j P_{ij} \log \frac{1}{P_{ij}} \right) p_i$$

---

[2] In Analysis, the Heine–Borel Theorem, which we do not need here, asserts that this is equivalent to the more general definition of compactness in terms of open sets.

is a linear function of $\mathbf{p}$, since each P_{ij} is constant, and hence it is a continuous function of $\mathbf{p}$. Similarly

$$H(\mathcal{B}) = \sum_j q_j \log \frac{1}{q_j}$$

is a continuous function of $\mathbf{q} = (q_j)$, because $x \log x$ is a continuous function of x on $[0,1]$, and $\mathbf{q} = \mathbf{p}M$ is a continuous function of $\mathbf{p}$, because $M = (P_{ij})$ is a constant matrix, so $H(\mathcal{B})$ is a continuous function of $\mathbf{p}$, since it is the composition

$$\mathbf{p} \mapsto \mathbf{p}M = \mathbf{q} = (q_j) \mapsto \sum_j q_j \log q_j = H(\mathcal{B})$$

of two continuous functions. Since $I(\mathcal{A}, \mathcal{B}) = H(\mathcal{B}) - H(\mathcal{B} \mid \mathcal{A})$, and the difference of two continuous functions is continuous, we have proved:

Theorem 4.14

The mutual information $I(\mathcal{A}, \mathcal{B})$ of each channel Γ is a continuous function of the input probability distribution vector $\mathbf{p} = (p_i)$.

Corollary 4.15

The mutual information $I(\mathcal{A}, \mathcal{B})$ of each channel Γ has a maximum value.

Proof

$I(\mathcal{A}, \mathcal{B})$ is a continuous function of $\mathbf{p}$, and $\mathbf{p}$ ranges over a compact set $\mathcal{P}$. A theorem in Analysis (see [La83], for instance) states that a continuous real-valued function on a compact set is bounded above and attains its least upper bound, so this is its maximum value.   □

This justifies our definition of the capacity C of Γ as this maximum value. In Chapter 5 we will show that C is also the least upper bound for the rates at which information can be transmitted accurately through Γ.

4.9 Supplementary Exercises

Exercise 4.12

Let n identical copies of a binary symmetric channel Γ be connected in series, as in Exercise 4.1. Show that the resulting channel is another binary symmetric channel, and calculate its capacity. (Hint: consider the eigenvalues of the channel matrices.) What happens as $n \to \infty$?

Exercise 4.13

Show that a channel has capacity $C = 0$ if and only if the rows of its channel matrix are all equal to each other. What is the interpretation of this, from the receiver's point of view?

Exercise 4.14

One can regard the entropy function $H(\mathbf{p}) = -\sum_i p_i \log p_i$ as a function $\mathcal{P} \to \mathbf{R}$, where $\mathcal{P}$ is the set of all probability distribution vectors $\mathbf{p} = (p_1, \ldots, p_r) \in \mathbf{R}^r$. Show that H is strictly convex on $\mathcal{P}$, in the sense that if $\mathbf{p}, \mathbf{q} \in \mathcal{P}$ and $\lambda \in [0, 1]$ then $H(\lambda \mathbf{p} + \bar{\lambda} \mathbf{q}) \geq \lambda H(\mathbf{p}) + \bar{\lambda} H(\mathbf{q})$, with equality if and only if $\mathbf{p} = \mathbf{q}$ or $\lambda = 0$ or 1.

Exercise 4.15

A channel Γ is *uniform* if each row of the channel matrix is a permutation of the entries $P_{11}, \ldots, P_{1s}$ in the first row, and similarly for the columns. Show that Γ has capacity

$$\log s + \sum_{j=1}^{s} P_{ij} \log P_{ij},$$

attained by an equiprobable input distribution. Hence find the capacity of the *r-ary symmetric channel*, for which $r = s$, $P_{ii} = P$ for all i, and $P_{ij} = \bar{P}/(r-1)$ for all $i \neq j$.

Exercise 4.16

The *general binary channel* Γ has a 2×2 channel matrix (P_{ij}), where $P_{i1} + P_{i2} = 1$ for $i = 1, 2$. Show that Γ has mutual information

$$I(\mathcal{A}, \mathcal{B}) = -q_1 \log q_1 - q_2 \log q_2 + q_1 c_1 + q_2 c_2,$$

where q_1, q_2 are the output probabilities, and c_1, c_2 are chosen so that $P_{i1}c_1 + P_{i2}c_2 = P_{i1} \log P_{i1} + P_{i2} \log P_{i2}$ for $i = 1, 2$. Deduce that Γ has capacity $C = \log(2^{c_1} + 2^{c_2})$. (Hint: use the technique of Lagrange multipliers to maximise $I(\mathcal{A}, \mathcal{B})$, subject to the constraint $q_1 + q_2 = 1$.) What happens when $P_{11} = P_{22}$? (This exercise is based on work of Muroga [Mu53].)

Exercise 4.17

Show that if channels Γ_1 and Γ_2 have capacities C_1 and C_2, their sum $\Gamma_1 + \Gamma_2$ has capacity $\log(2^{C_1} + 2^{C_2})$. How do you interpret this result when $\Gamma_1 = \Gamma_2$?

Exercise 4.18

Let Γ be a cascade $\Gamma_1 \circ \Gamma_2$ of channels, where Γ_1 has input $\mathcal{A}$ and output $\mathcal{B}$, and Γ_2 has input $\mathcal{B}$ and output $\mathcal{C}$. Show that

$$H(\mathcal{A} \mid \mathcal{C}) - H(\mathcal{A} \mid \mathcal{B}) =$$
$$\sum_b \sum_c \left(\Pr(b,c) \sum_a \Pr(a \mid b) \left(\log \Pr(a \mid b) - \log \Pr(a \mid c) \right) \right),$$

where the summations are over all the symbols a, b and c of $\mathcal{A}, \mathcal{B}$ and $\mathcal{C}$. Deduce that $H(\mathcal{A} \mid \mathcal{C}) \geq H(\mathcal{A} \mid \mathcal{B})$, and give an intuitive explanation of this result. Hence prove the *Data-processing Theorem* $I(\mathcal{A}, \mathcal{C}) \leq I(\mathcal{A}, \mathcal{B})$, which shows that mutual information cannot be increased by further transmission (a similar argument gives $I(\mathcal{A}, \mathcal{C}) \leq I(\mathcal{B}, \mathcal{C})$). Show that if each Γ_i has capacity C_i, then Γ has capacity $C \leq \min(C_1, C_2)$; give examples where $C = \min(C_1, C_2)$ and where $C < \min(C_1, C_2)$.

5

Using an Unreliable Channel

Let no such man be trusted. (*The Merchant of Venice*)

In this chapter, we assume that we are given an unreliable channel Γ, such as a BSC with $P < 1$, and that our task is to transmit information through Γ as accurately as possible. Shannon's Fundamental Theorem, which is perhaps the most important result in Information Theory, states that the capacity C of Γ is the least upper bound for the rates at which one can transmit information accurately through Γ. After first explaining some of the concepts involved, we will look at a simple example of how this accurate transmission might be achieved. A full proof of Shannon's Theorem is technically quite difficult, so for simplicity we will restrict the proof to the case where Γ is the BSC; we will give an outline proof for this channel in §5.4, postponing a more detailed proof to Appendix C.

5.1 Decision Rules

Let Γ be an information channel, with input $\mathcal{A}$ and output $\mathcal{B}$. The receiver, who sees each output symbol $b = b_j \in B$ emerging from Γ, needs an algorithm to decide which input symbol $a = a_i \in A$ gave rise to b_j. This will take the form of a *decision rule*, that is, a function $\Delta : B \rightarrow A$. Whenever b_j emerges from Γ, the receiver applies Δ to b_j, determines $a_i = \Delta(b_j)$, and decides (possibly incorrectly) that a_i was sent; we call this *decoding* the output. We will write

$i = j^*$ here, so that $\Delta(b_j) = a_{j^*}$.

The problem is that, in general, there are many functions $\Delta : B \to A$, and it may not be immediately clear which is the best one to use.

Exercise 5.1

How many different decision rules are there for a given information channel?

Example 5.1

Let Γ be the BSC, so that $A = B = \mathbf{Z}_2$. If the receiver trusts this channel, then Δ should be the identity function, that is, $\Delta(0) = 0$, $\Delta(1) = 1$; if not, another function $\Delta : \mathbf{Z}_2 \to \mathbf{Z}_2$ should be used (see Example 4.5 for a situation where it is reasonable to take $\Delta(0) = \Delta(1) = 0$).

If b_j is received, then the receiver decides that a_{j^*} was sent. The probability that this decision is correct is

$$\Pr\left(a = a_{j^*} \mid b = b_j\right) = Q_{j^*j} \, .$$

(See §4.1 for the definitions and the notation for probabilities used here.) Each b_j is received with probability q_j, so averaging over all $b_j \in B$, we see that the average probability $\Pr_C$ of correct decoding is given by

$$\Pr_C = \sum_j q_j Q_{j^*j} = \sum_j R_{j^*j} \, , \tag{5.1}$$

where $R_{ij} = q_j Q_{ij}$ is the joint probability $\Pr(a_i, b_j)$. It follows that the error-probability $\Pr_E$ (the average probability of incorrect decoding) is given by

$$\Pr_E = 1 - \Pr_C = 1 - \sum_j R_{j^*j} = \sum_j \sum_{i \neq j^*} R_{ij} \, . \tag{5.2}$$

Given Γ and $\mathcal{A}$, we want to choose a decision rule $\Delta : B \to A$ which minimises $\Pr_E$, or equivalently, which maximises $\Pr_C$; such a rule is sometimes called an *ideal observer* rule. For each j, we choose $i = j^*$ to maximise the backward probability $\Pr(a_i \mid b_j) = Q_{ij}$. (If there are several such i, we choose one of them arbitrarily as j^*.) This is equivalent to maximising the joint probability $R_{ij} = q_j Q_{ij}$ for each j, that is, $R_{j^*j} \geq R_{ij}$ for all i, so R_{j^*j} is the largest entry in column j of the matrix (R_{ij}). Since $R_{ij} = p_i P_{ij}$, this matrix can be found from the channel matrix $M = (P_{ij})$ and the input distribution (p_i) as

$$(R_{ij}) = \begin{pmatrix} p_1 & & \\ & \ddots & \\ & & p_r \end{pmatrix} M$$

(where the missing off-diagonal entries are all 0).

Example 5.2

If Γ is the BSC then by §4.2 we have

$$(R_{ij}) = \begin{pmatrix} p & 0 \\ 0 & \overline{p} \end{pmatrix} \begin{pmatrix} P & \overline{P} \\ \overline{P} & P \end{pmatrix} = \begin{pmatrix} pP & p\overline{P} \\ \overline{p}\,\overline{P} & \overline{p}P \end{pmatrix},$$

so we take

$$\Delta(0) = \begin{cases} 0 & \text{if } pP > \overline{p}\,\overline{P} \\ 1 & \text{if } pP < \overline{p}\,\overline{P}, \end{cases} \quad \text{and} \quad \Delta(1) = \begin{cases} 1 & \text{if } \overline{p}P > p\overline{P} \\ 0 & \text{if } \overline{p}P < p\overline{P}, \end{cases}$$

with an arbitrary choice of $\Delta(b)$ in case of equality.

Exercise 5.2

Calculate $\mathrm{Pr_E}$, where the channel Γ and the input $\mathcal{A}$ are as in Example 4.5 (a BSC with $P = 0.8$ and $p = 0.9$), and Δ is the ideal observer rule.

In some situations, the receiver may know how the channel behaves, but not the input, so that the forward probabilities P_{ij} are known, but not the input probabilities p_i. This means that the probabilities Q_{ij} and R_{ij} are unknown, and cannot therefore be used to choose the decision rule Δ. When this happens, the receiver has to base the choice of Δ on the probabilities P_{ij}, which depend only on Γ. The obvious method is, for each j, to choose $i = j^*$ to maximise P_{ij}, so that P_{j^*j} is the largest entry in column j of the channel matrix $M = (P_{ij})$. (As usual, if there are several such entries we choose one of them arbitrarily.) Such a rule Δ, defined by $P_{j^*j} \geq P_{ij}$ for all i, is called a *maximum likelihood rule*; as before, $\mathrm{Pr_C}$ and $\mathrm{Pr_E}$ are given by (5.1) and (5.2). If the r input symbols a_i are equiprobable, then for each j the forward probabilities P_{ij} are proportional to the joint probabilities $R_{ij} = p_i P_{ij} = P_{ij}/r$, so this maximum likelihood rule coincides with the ideal observer rule, which maximises $\mathrm{Pr_C}$. For other input distributions, this rule may not be the best (see Example 5.4 below); however, over all distributions it is the best in the sense that it maximises the multiple integral

$$\int_{\mathbf{p} \in \mathcal{P}} \mathrm{Pr_C}\, dp_1 \ldots dp_r \,,$$

where $\mathcal{P}$ denotes the set of all probability distribution vectors $\mathbf{p} = (p_1, \ldots, p_r) \in \mathbf{R}^r$. This result says that one's natural intuition is correct: if nothing is known about the input then the various possibilities balance out, and the maximum likelihood rule is the best one can hope for.

Exercise 5.3

Prove the above claim that, among all the decision rules for a given channel, the maximum likelihood rule maximises the integral of Pr_C over all inputs $\mathbf{p} \in \mathcal{P}$.

Example 5.3

Let us apply the maximum likelihood rule Δ to the BSC, where $P > \frac{1}{2}$ (so Γ is more reliable than unreliable). Then $P > \overline{P}$, so choosing the greatest entry in each column of the channel matrix

$$M = \begin{pmatrix} P & \overline{P} \\ \overline{P} & P \end{pmatrix},$$

we take $\Delta(0) = 0$ and $\Delta(1) = 1$. This gives

$$\mathrm{Pr}_C = pP + \overline{p}P = P \qquad \text{and} \qquad \mathrm{Pr}_E = p\overline{P} + \overline{p}\,\overline{P} = \overline{P}.$$

If $P < \frac{1}{2}$, on the other hand, we have $\overline{P} > P$, so $\Delta(0) = 1$ and $\Delta(1) = 0$, giving

$$\mathrm{Pr}_C = \overline{p}\,\overline{P} + p\overline{P} = \overline{P} \qquad \text{and} \qquad \mathrm{Pr}_E = \overline{p}P + pP = P.$$

Example 5.4

For a specific illustration, let us return to Example 4.5, where $P = 0.8$ and $p = 0.9$. As we saw in Example 5.3, the maximum likelihood rule gives $\Delta(0) = 0$ and $\Delta(1) = 1$, with $\mathrm{Pr}_C = P = 0.8$. However the ideal observer rule gives $\Delta(0) = \Delta(1) = 0$, with $\mathrm{Pr}_C = 0.9 > 0.8$ (see Exercise 5.2), so here the maximum likelihood rule is not the best choice.

Example 5.5

Let Γ be the binary erasure channel (BEC) in Example 4.2, with $P > 0$. Then the maximum likelihood rule gives $\Delta(0) = 0$, $\Delta(1) = 1$, and $\Delta(?) = 0$ or 1, say $\Delta(?) = 0$. It follows that if the input probabilities for 0 and 1 are p and $\overline{p}$, then

$$\mathrm{Pr}_C = pP + \overline{p}P + p\overline{P} = P + p\overline{P} \qquad \text{and} \qquad \mathrm{Pr}_E = \overline{p}.0 + p.0 + \overline{p}\,\overline{P} = \overline{p}\,\overline{P}.$$

5.2 An Example of Improved Reliability

Given an unreliable channel, how can we transmit information through it with greater reliability? Before considering this problem in general, let us look at a simple example. We take Γ to be the BSC with $1 > P > \frac{1}{2}$; for notational simplicity, let us define $Q = \overline{P} = 1 - P$, so the channel matrix is

$$M = \begin{pmatrix} P & Q \\ Q & P \end{pmatrix}.$$

Since $P > Q$, Example 5.3 shows that the maximum likelihood rule is given by $\Delta(0) = 0$, $\Delta(1) = 1$, with $\mathrm{Pr}_E = Q$. Let us also assume that the input symbols are equiprobable, that is, $p = \overline{p} = \frac{1}{2}$; then the mutual information $I(\mathcal{A}, \mathcal{B})$ attains its maximum value, which is the channel capacity $C = 1 - H(P)$ (see §4.7 and §4.8).

   If the error-probability $\mathrm{Pr}_E = Q$ is unacceptably high, let us try to reduce it by sending each input symbol $a = 0$ or 1 three times in succession. This means that we use the code

$$\mathcal{C} : 0 \mapsto 000, \ 1 \mapsto 111 \,,$$

so the input $\mathcal{C}$ now consists of two equiprobable words $w = 000$ and 111. During transmission through Γ, any of the three symbols in w could be changed, so the output $\mathcal{D}$ consists of eight binary words $000, 001, 010, 100, 011, 101, 110, 111$ of length 3. Now each symbol of w has probability P or Q of being correctly or incorrectly transmitted, so the forward probabilities for this new input and output are given by the matrix

$$\begin{pmatrix} P^3 & P^2Q & P^2Q & P^2Q & PQ^2 & PQ^2 & PQ^2 & Q^3 \\ Q^3 & PQ^2 & PQ^2 & PQ^2 & P^2Q & P^2Q & P^2Q & P^3 \end{pmatrix},$$

where the rows and columns correspond to the words in $\mathcal{C}$ and $\mathcal{D}$ in the stated order. Since $P > Q > 0$ we have $P^3 > Q^3$ and $P^2Q > PQ^2$, so the maximum likelihood rule is given by

$$\Delta : \begin{cases} 000, 001, 010, 100 \mapsto 000, \\ 011, 101, 110, 111 \mapsto 111. \end{cases}$$

By composing this with the decoding function (the inverse of $\mathcal{C}$)

$$000 \mapsto 0, \quad 111 \mapsto 1,$$

we can decode the words of $\mathcal{D}$ according to the rule

$$000, 001, 010, 100 \mapsto 0,$$
$$011, 101, 110, 111 \mapsto 1.$$

(This is sometimes called *majority decoding*: we count the symbols 0 and 1 in the received word, and take the most frequent. By using words of odd length, we can guarantee that one symbol will always have a clear majority over the other.)

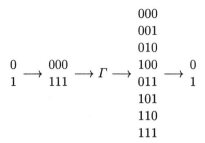

$$
\begin{array}{c}
000 \\
001 \\
010 \\
\begin{array}{ccc}
0 \\ 1
\end{array}
\longrightarrow
\begin{array}{c}
000 \\ 111
\end{array}
\longrightarrow \Gamma \longrightarrow
\begin{array}{c}
100 \\ 011
\end{array}
\longrightarrow
\begin{array}{c}
0 \\ 1
\end{array} \\
101 \\
110 \\
111
\end{array}
$$

Figure 5.1

The whole process of encoding, transmitting and decoding is summarised in Fig. 5.1. In effect, we have now constructed a new binary symmetric channel Γ'; an input symbol $a = 0$ or 1 is encoded by C as a word $w = 000$ or 111, which is transmitted through Γ; the received word is then decoded as an output symbol $b = 0$ or 1 by majority decoding. Now decoding is correct $(b = a)$ if and only if at most one of the three symbols in w is changed during transmission through Γ. Each symbol of w is correctly transmitted with probability P, so the probability of there being no errors in w is P^3. There are three ways in which a single error can occur, with one of the three symbols being transmitted incorrectly and the other two correctly; each of these three cases has probability P^2Q, so the probability of a single error is $3P^2Q$. Similarly, the probabilities of two and three errors are $3PQ^2$ and Q^3. Thus the channel matrix for Γ' is

$$
M' = \begin{pmatrix} P^3 + 3P^2Q & 3PQ^2 + Q^3 \\ 3PQ^2 + Q^3 & P^3 + 3P^2Q \end{pmatrix},
$$

so Γ' is a BSC with probability $P' = P^3 + 3P^2Q$ of correct transmission. The error-probability is therefore

$$
\mathrm{Pr_E} = \overline{P'} = 3PQ^2 + Q^3 = Q^2(3 - 2Q) \approx 3Q^2,
$$

which is significantly less than the original error-probability Q for small $Q > 0$. (For instance, if $Q = 0.01$ then $\mathrm{Pr_E} = 0.000298$.) Thus we have improved the error-probability, but at the cost of slower transmission: it now takes a code-word of length 3 to transmit a single input symbol, so we say that the rate of transmission is $R = 1/3$ (compared with its original value of 1).

There is an obvious extension of this idea, using code-words $00\ldots0$ and $11\ldots1$ of any length n to transmit the symbols 0 and 1; this is the *binary repetition code* $\mathcal{R}_n$ of length n. If we take n to be odd, then the maximum likelihood rule is majority decoding, as shown by Exercise 5.10 in §5.7. (If n is even, the received word might contain the same number of symbols 0 and 1, giving no majority.) One can show that $\mathrm{Pr_E} \to 0$ as $n \to \infty$ (see Exercise 5.10

again); the following table gives the approximate values of $\mathrm{Pr_E}$ for odd $n \leq 11$, on the assumption that $Q = 0.01$:

$n =$	1	3	5	7	9	11
$\mathrm{Pr_E} \approx$	10^{-2}	3×10^{-4}	10^{-5}	3.5×10^{-7}	1.3×10^{-8}	5×10^{-10}

However, the transmission rate $R = 1/n \to 0$ also, so we have bought increased accuracy at the cost of slower transmission.

This idea can be generalised further. If Γ is a channel with an input $\mathcal{A}$ having an alphabet A of r symbols, then any subset $\mathcal{C} \subseteq A^n$ can be used as a set of code-words which are transmitted through Γ. For instance, the repetition code $\mathcal{R}_n$ over A consists of all the words $w = aa \ldots a$ of length n such that $a \in A$; we will consider some further examples in Chapters 6 and 7. We call $\mathcal{C}$ an r-ary code of *length* n. If $|\mathcal{C}| = r^k$ then $\mathcal{C}$ can encode the k-th extension $\mathcal{A}^k$ (since this consists of r^k words); the n symbols of each code-word in $\mathcal{C}$ represent k symbols emitted by $\mathcal{A}$, so we say that $\mathcal{C}$ has rate $R = k/n$. For instance, the r-ary repetition code $\mathcal{R}_n$ has $|\mathcal{R}_n| = r$, so $k = 1$ and $R = 1/n$. More generally, the *rate* (or *transmission-rate*) of any non-empty code $\mathcal{C} \subseteq A^n$ is defined to be

$$R = \frac{\log_r |\mathcal{C}|}{n}, \tag{5.3}$$

so that $|\mathcal{C}| = r^{Rn}$. Since $|A^n| = r^n$, we see that $0 \leq R \leq 1$ for every code $\mathcal{C}$.

Shannon's Fundamental Theorem (which we shall consider in §5.4) states that, by choosing codes $\mathcal{C} \subseteq A^n$ for sufficiently large n, and by using suitable decision rules, we can make the error-probability $\mathrm{Pr_E}$ approach 0, *without* the transmission-rate R also approaching 0 (as it does for $\mathcal{R}_n$); in fact, we can do this with the rate R arbitrarily close to the channel capacity C.

5.3 Hamming Distance

The previous section illustrated a simple example of how to transmit information with improved accuracy. One important ingredient in the method was the choice of code-words $00 \ldots 0$ and $11 \ldots 1$ which are very different from each other, so that even if they are received with errors, the receiver is still likely to be able to distinguish them. When we try to extend this idea to construct more effective codes, we will need to choose larger sets of code-words which are also very unlike each other. To measure how like or unlike each other two words are, we introduce a notion of distance between words.

Let $\mathbf{u} = u_1 \ldots u_n$ and $\mathbf{v} = v_1 \ldots v_n$ be words of length n in some alphabet A, so $\mathbf{u}, \mathbf{v} \in A^n$. (We write these words in bold-face because we will soon need to regard them as vectors $(u_1, \ldots, u_n)$ in a vector space A^n, where A is a field.)

The *Hamming distance* $d(\mathbf{u}, \mathbf{v})$ between $\mathbf{u}$ and $\mathbf{v}$ is defined to be the number of subscripts i such that $u_i \neq v_i$.

Example 5.6

Let $\mathbf{u} = 01101$ and $\mathbf{v} = 01000$ in $\mathbf{Z}_2^5$. Then $d(\mathbf{u}, \mathbf{v}) = 2$, since the words $\mathbf{u}$ and $\mathbf{v}$ differ in two positions ($i = 3$ and 5).

> ### Exercise 5.4
>
> If $\mathbf{u} \in A^n$ where $|A| = r$, and $0 \leq i \leq n$, then how many words $\mathbf{v} \in A^n$ have Hamming distance $d(\mathbf{u}, \mathbf{v}) = i$? Check that these numbers, for $i = 0, 1, \ldots, n$, add up to $|A^n|$.

Example 5.7

We can regard the words in $\mathbf{Z}_2^3$ as the eight vertices of a cube, as shown in Fig. 5.2.

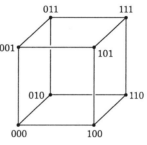

Figure 5.2

Then $d(\mathbf{u}, \mathbf{v})$ is not the euclidean distance, but rather the least number of edges in any path along the edges between $\mathbf{u}$ and $\mathbf{v}$. (This notion of distance is used in Graph Theory, where the distance between two vertices in a connected graph is defined to be the least number of edges in any path from one vertex to the other.)

> ### Exercise 5.5
>
> How large can a subset $\mathcal{C} \subseteq \mathbf{Z}_2^3$ be, if $d(\mathbf{u}, \mathbf{v}) \geq 2$ for all $\mathbf{u} \neq \mathbf{v}$ in $\mathcal{C}$? Describe geometrically all the subsets attaining this bound. What is the analogous bound for subsets of $\mathbf{Z}_2^n$?

Lemma 5.8

Let $\mathbf{u}, \mathbf{v}, \mathbf{w} \in A^n$. Then

(a) $d(\mathbf{u}, \mathbf{v}) \geq 0$, with equality if and only if $\mathbf{u} = \mathbf{v}$;

(b) $d(\mathbf{u}, \mathbf{v}) = d(\mathbf{v}, \mathbf{u})$;

(c) $d(\mathbf{u}, \mathbf{w}) \leq d(\mathbf{u}, \mathbf{v}) + d(\mathbf{v}, \mathbf{w})$.

Proof

(a) Obvious. (b) Trivial. (c) Easy (see Exercise 5.6). $\qquad\square$

Part (c) is known as the *triangle inequality*, since it expresses the fact that one side **uw** of a triangle **uvw** cannot have length greater than the sum of the lengths of the other two sides **uv** and **vw**.

Exercise 5.6

Prove Lemma 5.8(c).

In Topology, a set with a function d satisfying conditions (a), (b) and (c) is known as a *metric space*, though we will not use this fact. The point of this result is to show that the Hamming distance d behaves very much like the euclidean distance-function in $\mathbf{R}^n$.

To transmit information through Γ, we choose a code $\mathcal{C} \subseteq A^n$ for some n, and use the maximum likelihood decision rule (§5.1): we decode each received word as the code-word most likely to have caused it. This is the best decision rule if the code-words are equiprobable, and it is the best rule in general if the probabilities are unknown. Even if, for some particular probability distribution of code-words, it is not the best decision rule, it is good enough to give $\mathrm{Pr_E} \to 0$ as $n \to \infty$ in the proof of Shannon's Theorem.

For simplicity, we will assume for the rest of this section and the next that Γ is the BSC, with $P > \frac{1}{2}$, so $A = B = \mathbf{Z}_2$ and $r = 2$. Our choice of the maximum likelihood decision rule means that, for any output $\mathbf{v} \in \mathbf{Z}_2^n$, we decode $\mathbf{v}$ as the code-word $\mathbf{u} = \Delta(\mathbf{v}) \in \mathcal{C}$ which maximises the forward probability $\mathrm{Pr}\,(\mathbf{v} \mid \mathbf{u})$. Now if $d(\mathbf{u}, \mathbf{v}) = i$ then

$$\mathrm{Pr}\,(\mathbf{v} \mid \mathbf{u}) = Q^i P^{n-i},$$

the probability of errors in the i places where **u** and **v** differ, and of correct transmission in the remaining $n - i$ places. Thus

$$\mathrm{Pr}\,(\mathbf{v} \mid \mathbf{u}) = P^n \left(\frac{Q}{P}\right)^i,$$

which is a decreasing function of i since $Q/P < 1$, so a code-word **u** which maximises $\mathrm{Pr}\,(\mathbf{v} \mid \mathbf{u})$ is one which minimises $d(\mathbf{u}, \mathbf{v})$. Thus the maximum likelihood rule Δ decodes each received word $\mathbf{v} \in \mathbf{Z}_2^n$ as the code-word $\mathbf{u} = \Delta(\mathbf{v}) \in \mathcal{C}$ which is closest to **v** with respect to the Hamming distance. This rule, which we will use for the rest of Chapter 5, is called *nearest neighbour decoding*. As usual, we make an arbitrary choice of nearest neighbour if there there are two or more of them.

5.4 Statement and Outline Proof of Shannon's Theorem

The following result, often called the Fundamental Theorem of Information Theory, was proved by Shannon [Sh48] in 1948. Informally, it says that if we use long enough code-words then we can send information through a channel Γ as accurately as we require, at a rate arbitrarily close to the capacity C of Γ. This is an improvement on the previous example of the repetition code $\mathcal{R}_n$, which provides the desired accuracy as $n \to \infty$, but which has rate approaching 0 rather than C. For simplicity we shall state and prove the theorem for the BSC, but in fact it is valid for all channels. The precise statement is as follows:

Theorem 5.9

Let Γ be a binary symmetric channel with $P > \frac{1}{2}$, so Γ has capacity $C = 1 - H(P) > 0$, and let δ, $\varepsilon > 0$. Then for all sufficiently large n there is a code $\mathcal{C} \subseteq \mathbf{Z}_2^n$, of rate R satisfying $C - \varepsilon \le R < C$, such that nearest neighbour decoding gives error-probability $\mathrm{Pr_E} < \delta$.

Thus, by taking δ and ε sufficiently small, we can make $\mathrm{Pr_E}$ and R as close as required to 0 and C respectively. A complete proof, which is rather long and involved, is given in Appendix C. Here we simply sketch the main ideas.

Outline Proof

Let $R < C$ (as close as we like), and for large n randomly choose a set $\mathcal{C}$ consisting of 2^{nR} words in $\mathbf{Z}_2^n$. (We will ignore the inconvenient possibility that 2^{nR} may not be an integer! In a rigorous proof, we choose an integer close to 2^{nR}, and show that this small adjustment does not affect the result.) This gives us a binary code $\mathcal{C}$ of length n; by (5.3) it has rate $\log_2(2^{nR})/n = R$.

If a code-word $\mathbf{u} \in \mathcal{C}$ is transmitted through Γ, then each of the n symbols in $\mathbf{u}$ has probability $\overline{P} = Q$ of error, so we expect about nQ of the symbols to be transmitted incorrectly. In fact, the Law of Large Numbers (see Appendix B) implies that this will happen with probability approaching 1 as $n \to \infty$. This means that we should expect the received word $\mathbf{v}$ to satisfy $d(\mathbf{u}, \mathbf{v}) \approx nQ$. Equivalently, from the receiver's point of view, if a word $\mathbf{v}$ is received, then it probably came from a code-word $\mathbf{u} \in \mathcal{C}$ satisfying $d(\mathbf{u}, \mathbf{v}) \approx nQ$.

The nearest neighbour rule decodes each received word $\mathbf{v}$ as the code-word $\Delta(\mathbf{v}) \in \mathcal{C}$ closest to $\mathbf{v}$, so if decoding is incorrect then there must be some $\mathbf{u}' \ne \mathbf{u}$ in $\mathcal{C}$ with $d(\mathbf{u}', \mathbf{v}) \le d(\mathbf{u}, \mathbf{v})$. It follows that the probability of incorrectly decoding $\mathbf{v}$ is no greater than the probability that such a code-word $\mathbf{u}'$ exists,

so

$$\text{Pr}_E \leq \sum_{\mathbf{u}' \neq \mathbf{u}} \text{Pr}\left(d(\mathbf{u}', \mathbf{v}) \leq nQ\right), \tag{5.4}$$

where we have replaced $d(\mathbf{u}, \mathbf{v})$ with its approximate value nQ, and have ignored the small change in probability resulting from this. Since there are $|\mathcal{C}| - 1 = 2^{nR} - 1$ code-words $\mathbf{u}' \neq \mathbf{u}$, and they are randomly chosen, the upper bound on Pr_E in (5.4) is equal to

$$(|\mathcal{C}| - 1)\,\text{Pr}\left(d(\mathbf{u}', \mathbf{v}) \leq nQ\right) < 2^{nR}\,\text{Pr}\left(d(\mathbf{u}', \mathbf{v}) \leq nQ\right).$$

Now we chose the code-words $\mathbf{u}'$ randomly from $\mathbf{Z}_2^n$, so for any given $\mathbf{v}$, the probability that $d(\mathbf{u}', \mathbf{v}) \leq nQ$ is equal to the proportion of the 2^n words $\mathbf{u}' \in \mathbf{Z}_2^n$ satisfying this inequality. For any given $\mathbf{v}$ and i, the number of words $\mathbf{u}' \in \mathbf{Z}_2^n$ satisfying $d(\mathbf{u}', \mathbf{v}) = i$ is equal to the binomial coefficient $\binom{n}{i}$, the number of ways of choosing i of the n symbols in $\mathbf{v}$ to be different in $\mathbf{u}'$. This implies that the number of words $\mathbf{u}' \in \mathbf{Z}_2^n$ satisfying $d(\mathbf{u}', \mathbf{v}) \leq nQ$ is $\sum_{i \leq nQ} \binom{n}{i}$, so

$$\text{Pr}\left(d(\mathbf{u}', \mathbf{v}) \leq nQ\right) = \frac{1}{2^n} \sum_{i \leq nQ} \binom{n}{i}.$$

To continue the proof, we need the following result, which is also used in the complete proof of Shannon's Theorem in Appendix C.

Exercise 5.7

Show that if $\lambda + \mu = 1$, where $0 \leq \lambda \leq \frac{1}{2}$, then

$$1 \geq \sum_{i \leq \lambda n} \binom{n}{i} \lambda^i \mu^{n-i} \geq \sum_{i \leq \lambda n} \binom{n}{i} \lambda^{\lambda n} \mu^{\mu n} ;$$

hence show that

$$\sum_{i \leq \lambda n} \binom{n}{i} \leq 2^{nH(\lambda)}.$$

(Compare this with the well-known identity $\sum_{i=0}^{n} \binom{n}{i} = 2^n$.)

Continuation of Proof

We now return to the proof of Theorem 5.9. Putting $\lambda = Q$ in Exercise 5.7 we have

$$\sum_{i \leq nQ} \binom{n}{i} \leq 2^{nH(Q)};$$

thus (5.4) becomes

$$\text{Pr}_E < 2^{nR} \cdot \frac{1}{2^n} \cdot 2^{nH(Q)} = 2^{n(R-1+H(Q))} = 2^{n(R-C)},$$

since Γ has capacity $C = 1 - H(P) = 1 - H(Q)$ (see §4.8). Now $R < C$, so $2^{n(R-C)} \to 0$ as $n \to \infty$, and hence $\mathrm{Pr_E} \to 0$ also.   □

Warning

You will probably have noticed several gaps in this proof, and unless you have read it very carefully, there may be others you have not noticed: for instance, in the final sentence we need to ensure that $R - C$ is bounded away from 0 as $n \to \infty$, so that $n(R - C) \to -\infty$. Nevertheless, it is possible to give a completely rigorous proof, based on the above outline. Since the remaining chapters do not depend on this proof, we have placed it in Appendix C to avoid interrupting the flow of ideas.

5.5 The Converse of Shannon's Theorem

Shannon's Theorem states that one can transmit information through Γ, as accurately as required, at rates $R < C$ which are arbitrarily close to the capacity C. An obvious question is whether one can do better than this, that is, whether one can replace C here with some constant $C' > C$. In this section we will show that this is impossible, so C is the least upper bound of the rates at which transmission of arbitrary accuracy is possible. We will prove this, not just for the BSC, but for arbitrary channels. First we need the *Fano bound*, which gives a lower bound on the error-probability.

Theorem 5.10

Let Γ be a channel with input $\mathcal{A}$ and output $\mathcal{B}$. Then the error-probability $\mathrm{Pr_E}$ corresponding to any decision rule Δ for Γ satisfies

$$H(\mathcal{A} \mid \mathcal{B}) \le H(\mathrm{Pr_E}) + \mathrm{Pr_E} \log(r - 1), \tag{5.5}$$

where r is the number of symbols in $\mathcal{A}$.

The values of $\mathrm{Pr_E}$ and $H(\mathcal{A} \mid \mathcal{B})$ satisfying this inequality are indicated by the shaded region in Fig. 5.3. Before proving Theorem 5.10, let us try to interpret this result. The receiver, knowing the output symbol b_j, uses a decision rule Δ to find $a_{j^*} = \Delta(b_j)$, which may or may not be the actual symbol a_i transmitted. The left-hand side of (5.5) is the equivocation of Γ, the extra information the receiver needs (on average) in order to know a_i. This extra information can be divided into two parts:

(a) whether or not decoding is correct, that is, whether or not $a_{j^*} = a_i$;

(b) if decoding is incorrect, then which a_i ($i \ne j^*$) was transmitted.

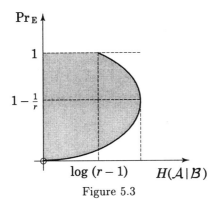

Figure 5.3

Now decoding is incorrect or correct, with probabilities $\mathrm{Pr_E}$ and $\mathrm{Pr_C} = \overline{\mathrm{Pr_E}}$, so the information in (a) has value $H(\mathrm{Pr_E})$, the first term on the right-hand side of (5.5). If decoding is correct, the receiver needs no more information, but if it is incorrect (which happens with probability $\mathrm{Pr_E}$) then the receiver needs to know which of the $r - 1$ symbols $a_i \neq a_{j^*}$ was transmitted. This involves specifying one symbol out of $r - 1$, so by putting $q = r - 1$ in Theorem 3.10 we see that this information is worth at most $\log(r - 1)$. Multiplying this by the probability that (b) is needed, we find that on average the information in (b) is worth at most $\mathrm{Pr_E} \log(r - 1)$, the second term on the right in (5.5).

Proof of Theorem 5.10

Without loss of generality, we can use entropies and logarithms with base r. In §4.3 we saw that

$$H(\mathcal{A} \mid \mathcal{B}) = \sum_i \sum_j R_{ij} \log \frac{1}{Q_{ij}} = -\sum_j \sum_i R_{ij} \log Q_{ij}.$$

We split the terms in this double summation into two sets, corresponding to correct and incorrect decoding ($i = j^*$ and $i \neq j^*$ respectively):

$$H(\mathcal{A} \mid \mathcal{B}) = -\sum_j R_{j^*j} \log Q_{j^*j} - \sum_j \sum_{i \neq j^*} R_{ij} \log Q_{ij}.$$

Now

$$H(\mathrm{Pr_E}) = -\mathrm{Pr_E} \log \mathrm{Pr_E} - \mathrm{Pr_C} \log \mathrm{Pr_C},$$

where

$$\mathrm{Pr_C} = \sum_j R_{j^*j} \quad \text{and} \quad \mathrm{Pr_E} = \sum_j \sum_{i \neq j^*} R_{ij},$$

so

$$H(\mathcal{A} \mid \mathcal{B}) - H(\mathrm{Pr_E}) - \mathrm{Pr_E} \log(r - 1) =$$

$$\sum_j R_{j^*j} \log \frac{\mathrm{Pr_C}}{Q_{j^*j}} + \sum_j \sum_{i \neq j^*} R_{ij} \log \frac{\mathrm{Pr_E}}{(r - 1)Q_{ij}} \, .$$

The base r version of Lemma 3.8 gives $\log x \leq (x - 1) \log e$ for all $x > 0$, so the right-hand side of this equation is

$$\leq \log e \left(\sum_j R_{j^*j} \left(\frac{\mathrm{Pr_C}}{Q_{j^*j}} - 1 \right) + \sum_j \sum_{i \neq j^*} R_{ij} \left(\frac{\mathrm{Pr_E}}{(r - 1)Q_{ij}} - 1 \right) \right)$$

$$= \log e \left(\mathrm{Pr_C} \sum_j q_j - \sum_j R_{j^*j} + \frac{\mathrm{Pr_E}}{(r - 1)} \sum_j \sum_{i \neq j^*} q_j - \sum_j \sum_{i \neq j^*} R_{ij} \right)$$

$$= \log e \left(\mathrm{Pr_C} - \mathrm{Pr_C} + \frac{\mathrm{Pr_E}}{(r - 1)}(r - 1) - \mathrm{Pr_E} \right)$$

$$= 0,$$

where we have used $R_{ij} = q_j Q_{ij}$ in the second line, and $\sum_j q_j = 1$ and $\sum_j \sum_i q_j = r$ in the third.   □

We can now prove what is sometimes known as the converse of Shannon's Theorem, namely that if $C' > C$ then it is not true that for every $\varepsilon > 0$ there is a sequence of codes $\mathcal{C}$, of lengths $n \to \infty$, and of rates R satisfying $C' - \varepsilon \leq R < C'$, such that $\mathrm{Pr_E} \to 0$ as $n \to \infty$. To prove this, it is sufficient to show that for some $\varepsilon > 0$ there is no such sequence of codes. Let us take $\varepsilon = (C' - C)/2$, and suppose that such a sequence of codes $\mathcal{C}$ exists. Since $C' - \varepsilon \leq R < C'$, each code has rate $R \geq C + \varepsilon$.

We can regard $\mathcal{C}$ as an input for the n-th extension Γ^n of Γ (as defined in §4.1), with output $\mathcal{D} = \mathcal{B}^n$. By applying the Fano bound (Theorem 5.10) to the channel Γ^n we see that

$$H(\mathcal{C} \mid \mathcal{D}) \leq H(\mathrm{Pr_E}) + \mathrm{Pr_E} \log(M - 1),$$

where $M = |\mathcal{C}|$. Alternatively, we can regard $\mathcal{C}$ as defining an input probability distribution on the input alphabet A^n of Γ^n, each of the $M = |\mathcal{C}|$ code-words in $\mathcal{C}$ having probability $1/M$, and all other words in A^n having probability 0. By Exercise 4.10, Γ^n has capacity nC, so

$$H(\mathcal{C}) - H(\mathcal{C} \mid \mathcal{D}) = I(\mathcal{C}, \mathcal{D}) \leq nC,$$

and hence

$$H(\mathcal{C}) - nC \le H(\mathcal{C} \mid \mathcal{D})$$
$$\le H(\mathrm{Pr_E}) + \mathrm{Pr_E}\log(M - 1)$$
$$\le H(\mathrm{Pr_E}) + \mathrm{Pr_E}\log M$$
$$= H(\mathrm{Pr_E}) + \mathrm{Pr_E}.nR$$

since $M = r^{nR}$. Now $\mathcal{C}$ has M equiprobable code-words, so $H(\mathcal{C}) = \log M = nR$ by Theorem 3.10, giving

$$nR - nC \le H(\mathrm{Pr_E}) + \mathrm{Pr_E}.nR.$$

Thus

$$0 < \varepsilon \le R - C \le \frac{1}{n}H(\mathrm{Pr_E}) + \mathrm{Pr_E}.R \le \frac{1}{n}H(\mathrm{Pr_E}) + \mathrm{Pr_E}$$

for all n, where ε is independent of n. However, if $\mathrm{Pr_E} \to 0$ then $H(\mathrm{Pr_E}) \to 0$ also, so the right-hand side is less than ε for all sufficiently large n. We have shown that this is false, so $\mathrm{Pr_E} \nrightarrow 0$ as $n \to \infty$. (In fact, a more careful argument shows that $\mathrm{Pr_E} \to 1$.)

Example 5.11

Let Γ be the BSC, and as a rather extreme example of a code let us take $\mathcal{C} = \mathcal{A}^n$, so $R = 1$. If $0 < P < 1$ we have $C = 1 - H(P) < 1$, so $R > C$. Using the identity function $\Delta(\mathbf{u}) = \mathbf{u}$ as a decision rule, we see that decoding is correct if and only if there are no errors, so $\mathrm{Pr_E} = 1 - P^n \to 1$ as $n \to \infty$.

Example 5.12

In §7.4 we will construct a sequence of binary codes $\mathcal{H}_n$ (the Hamming codes) of length n of the form $2^c - 1$ and rate $R = (n - c)/n$, so $R \to 1$ as $n \to \infty$. If we use a BSC with $0 < P < 1$, then $C = 1 - H(P) < 1$ and hence $R > C$ for all sufficiently large n. As we shall see in §7.4, nearest neighbour decoding is correct if and only if there is at most one error, so $\mathrm{Pr_E} = 1 - P^n - nP^{n-1}Q \to 1$ as $n \to \infty$.

5.6 Comments on Shannon's Theorem

The general form of Shannon's Theorem is as follows:

Theorem 5.13

Let Γ be an information channel with capacity $C > 0$, and let $\delta, \varepsilon > 0$. For all sufficiently large n there is a code $\mathcal{C}$ of length n, of rate R satisfying $C - \varepsilon \le R < C$, together with a decision rule which has error-probability $\mathrm{Pr_E} < \delta$.

The basic principles of the proof are similar to those for the BSC; see [As65] for the full details. Although this is a very powerful result, it has several limitations:

Comment 5.14

In order to achieve values of R close to C and $\mathrm{Pr_E}$ close to 0, one may have to use a very large value of n. This means that code-words are very long, so encoding and decoding may become difficult and time-consuming. Moreover, if n is large then the receiver experiences delays while waiting for complete code-words to come through; when a received word is decoded, there is a sudden burst of information, which may be difficult to handle.

Comment 5.15

Shannon's Theorem tells us that good codes exist, but neither the statement nor the proof give one much help in finding them. The proof shows that the "average" code is good, but there is no guarantee that any specific code is good: this has to be proved by examining that code in detail. One might choose a code at random, as in the proof of the Theorem, and there is a reasonable chance that it will be good. However, random codes are very difficult to use: ideally, one wants a code to have plenty of structure, which can then be used to design effective algorithms for encoding and decoding. We will see examples of this in Chapters 6 and 7, when we construct specific codes with good transmission-rates or error-probabilities.

5.7 Supplementary Exercises

Exercise 5.8

Let Γ be the BEC, with $P > 0$, and let the input probabilities be $p, \bar{p}$ with $0 < p < 1$. Show how to use the binary repetition code $\mathcal{R}_n$ to send information through Γ so that $\mathrm{Pr_E} \to 0$ as $n \to \infty$.

Exercise 5.9

A binary channel Γ always transmits 0 correctly, but transmits 1 as 1 or 0 with probabilities P and $Q = \bar{P}$, where $0 < P < 1$. Write down the channel matrix, and describe the maximum likelihood rule. If the input probabilities of 0 and 1 are p and $\bar{p}$, find $\mathrm{Pr_E}$. To improve reliability, 0 and 1 are encoded as 000 and 111. Describe the resulting maximum likelihood rule; is it the same as (i) majority decoding, (ii) nearest neighbour decoding? Find the resulting rate and error-probability. What happens if instead we use the binary repetition code $\mathcal{R}_n$, and let $n \to \infty$?

Exercise 5.10

The binary repetition code $\mathcal{R}_n$, of odd length $n = 2t + 1$, is used to encode messages transmitted through a BSC Γ in which each digit has probabilities P and Q $(= \overline{P})$ of correct or incorrect transmission, and $P > \frac{1}{2}$. Show that in this case the maximum likelihood rule, majority decoding and nearest neighbour decoding all give the same decision rule Δ. Show that this rule has error-probability

$$\mathrm{Pr_E} \leq \frac{(2t+1)!}{(t!)^2} P^t Q^{t+1},$$

and deduce that $\mathrm{Pr_E} \to 0$ as $n \to \infty$. Why does this not give a direct proof of Shannon's Fundamental Theorem?

Exercise 5.11

(This exercise and the next are based on work by Kelley [Ke56].) A gambler bets on the outcomes of a sequence of tosses of an unbiased coin, placing his bet after the coin is tossed, but before the outcome is announced. A correct bet wins back twice the stake, but an incorrect bet loses it. He decides to cheat by learning the outcome of each toss through a BSC Γ with probabilities P, Q of correct and incorrect transmission, then betting a fixed proportion λ of his capital on the symbol emitted by Γ, and the remaining $\mu = \overline{\lambda}$ on the other symbol. Show that if his initial capital is c_0, then after n tosses it is $c_n = 2^n \lambda^m \mu^{n-m} c_0$, where m is the number of times Γ gives correct information. Show that, over a long period, the exponential growth rate

$$G = \lim_{n \to \infty} \frac{1}{n} \log\left(\frac{c_n}{c_0}\right)$$

of the gambler's capital is probably given by

$$G \approx 1 + P \log \lambda + Q \log \mu.$$

Show that this is maximised by taking $\lambda = P$, in which case $G \approx C$, the capacity of Γ. If $\frac{1}{2} < P < 1$, how could the gambler benefit from reading this chapter?

Exercise 5.12

How does Exercise 5.11 generalise to the case where the gambler is the receiver of an arbitrary channel Γ, betting on the input symbols, and a successful bet on a symbol a_i of probability p_i regains $1/p_i$ times the stake? What would happen if we changed the odds (but not the probabilities p_i), so that a successful bet on a_i regained $1/p_i'$ times the stake, where $\sum_i p_i' = 1$, $p_i' > 0$? Would the gambler gain or lose from this?

6
Error-correcting Codes

To thine own self be true. (*Hamlet*)

Our aim now is to construct codes $\mathcal{C}$ with good transmission-rates R and low error-probabilities $\mathrm{Pr_E}$, as promised by Shannon's Fundamental Theorem (§5.4). This part of the subject goes under the name of Coding Theory (or Error-correcting Codes), as opposed to Information Theory, which covers the topics considered earlier. The construction of such codes is quite a difficult task, and we will concentrate on a few simple examples to illustrate some of the methods used to construct more advanced codes.

6.1 Introductory Concepts

We will assume from now on that we are using a channel Γ in which the input and output alphabets A and B are equal, as in the case of the BSC; there is no loss of generality in doing this, since if not we can always replace A and B with the common alphabet $A \cup B$. We will denote this common finite alphabet by F, since we will often choose it to be a field, so that we can use techniques from Algebra. In order to be a field, F must be closed under addition, subtraction, multiplication and division by non-zero elements, with the usual axioms such as $ab = ba$, $a(b+c) = ab+ac$, etc. Standard examples include the fields $\mathbf{Q}$, $\mathbf{R}$ and $\mathbf{C}$ of rational, real and complex numbers. These are infinite fields, but for our purposes we need to use *finite* fields, such as the field $\mathbf{Z}_p$ of integers mod (p), where p is prime. The basic result we need about finite fields is:

Theorem 6.1

(a) There is a finite field of order q if and only if $q = p^e$ for some prime p and integer $e \geq 1$.

(b) Any two finite fields of the same order are isomorphic.

Many Algebra textbooks (such as [KR83]) prove this result, so we will assume it without proof. The essentially unique field of order q is known as the *Galois field* F_q or $GF(q)$. If $e = 1$, so $q = p$ is prime, then $F_q = F_p = \mathbf{Z}_p$, the field of integers mod (p). However, if $e > 1$, so q is composite, then $\mathbf{Z}_q$ is *not* a field: for instance $p^e = 0$ in $\mathbf{Z}_q$, even though $p \neq 0$, so p is a zero-divisor. This means that $F_q \neq \mathbf{Z}_q$ for $e > 1$; instead one can define F_q to be the field obtained by adjoining to $\mathbf{Z}_p$ a root α of an irreducible polynomial $f(x)$ of degree e, just as the complex field $\mathbf{C}$ is obtained from $\mathbf{R}$ by adjoining the root $i = \sqrt{-1}$ of $f(x) = x^2 + 1$. The elements of F_q then have the form $a_0 + a_1\alpha + \cdots + a_{e-1}\alpha^{e-1}$ where $a_0, a_1, \ldots, a_{e-1} \in \mathbf{Z}_p$, with the obvious operations of addition and subtraction; the product of two such elements can be put into this form by using the equation $f(\alpha) = 0$ to reduce powers of α. We need $f(x)$ to be irreducible to avoid zero-divisors in F_q.

Example 6.2

The quadratic polynomial $f(x) = x^2 + x + 1$ has no roots in the field $\mathbf{Z}_2$ (since $f(0) = f(1) = 1$), so it has no linear factors and is therefore irreducible over $\mathbf{Z}_2$. If we adjoin a root α of $f(x)$ to $\mathbf{Z}_2$, we obtain a field

$$F_4 = \{a + b\alpha \mid a, b \in \mathbf{Z}_2\} = \{0, 1, \alpha, 1 + \alpha\}$$

of order $q = 4$, in which $\alpha^2 + \alpha + 1 = 0$, so that $\alpha^2 = -1 - \alpha = 1 + \alpha$. For instance, $\alpha(1 + \alpha) = \alpha + \alpha^2 = 1 + 2\alpha = 1$, so α and $1 + \alpha$ are multiplicative inverses of each other in F_4. See Supplementary Exercises 6.16 and 6.17 for similar constructions of finite fields.

For our purposes, the precise structure of F_q is usually unimportant, and it is sufficient simply to know that it exists for each prime-power q. However, there are more advanced codes, beyond the scope of this book, which depend on a deeper knowledge of finite fields. Arithmetic in F_q is similar to that in any other field, except that if $q = p^e$ then $p = 0$ in F_q; also, there is no natural order relation $<$ in F_q, as there is in $\mathbf{R}$ and $\mathbf{Q}$ but not in $\mathbf{C}$. In many cases we will concentrate on binary codes, so that $F_q = \mathbf{Z}_2 = \{0, 1\}$, with $1 + 1 = 0$.

From now on we will follow Shannon's Fundamental Theorem and use block codes, those in which all the code-words have the same length. This does not conflict with our earlier use of variable-length codes for efficiency: we can use such a code first, and then break the resulting code-sequence into successive blocks of the same length k, which we represent as code-words of a fixed length n. We try to choose these code-words to be as far apart as possible (with respect to the Hamming distance), so that the resulting code has good error-correcting properties.

If we use code-words of length n, then a code C of length n is a subset of the set $V = F^n$ of all n-tuples with coordinates in F. If F is a field then V is an n-dimensional vector space over F, in which the operations are componentwise addition and scalar multiplication: if $\mathbf{u} = u_1 \ldots u_n$, $\mathbf{v} = v_1 \ldots v_n \in V$ and $a, b \in F$ then $a\mathbf{u} + b\mathbf{v}$ is the word, or vector, with i-th component $au_i + bv_i$ for $i = 1, \ldots, n$. We say that C is a *linear* code (or a group code) if C is a linear subspace of V; this means that C is non-empty, and if $\mathbf{u}, \mathbf{v} \in C$ then $a\mathbf{u} + b\mathbf{v} \in C$ for all $a, b \in F$. In particular, every linear code contains the zero vector $\mathbf{0} = 00 \ldots 0$, since $\mathbf{0} = 0\mathbf{u} + 0\mathbf{v}$ for any $\mathbf{u}, \mathbf{v} \in C$.

Exercise 6.1

Prove that if C and C' are linear codes contained in V, then the codes $C \cap C'$ and $C + C' = \{\mathbf{u} + \mathbf{u}' \mid \mathbf{u} \in C, \mathbf{u}' \in C'\}$ are also linear. Under what circumstances is the code $C \cup C'$ linear?

Most codes are non-linear, in the sense that comparitively few subsets of V are linear subspaces; however, most of the codes currently studied and used are linear, because these are easier to understand and to use. One can prove an analogue of Shannon's Fundamental Theorem for linear codes: instead of choosing a random code $C \subseteq V$, as in the proof of Theorem 5.9, we choose a random subset of V as a basis for a linear code $C \subseteq V$, and then show that C has the required properties as $n \to \infty$.

We will always denote $|C|$ by M. When C is linear we have $M = q^k$, where $k = \dim(C)$ is the dimension of the subspace C; this is because each element of C has a unique expression $a_1\mathbf{u}_1 + \cdots + a_k\mathbf{u}_k$ where $a_1, \ldots, a_k \in F$ and $\mathbf{u}_1, \ldots, \mathbf{u}_k$ is a basis for C, and there are $|F| = q$ independent choices for each a_i. We then call C a linear $[n, k]$-code.

The rate of a code C is

$$R = \frac{\log_q M}{n},\qquad(6.1)$$

so in the case of a linear $[n, k]$-code we have

$$R = \frac{k}{n}.\qquad(6.2)$$

We can interpret this by regarding k of the n digits in each code-word as information digits, carrying the information we wish to transmit, and the remaining $n - k$ as check digits, confirming or protecting that information.

From now onwards, we will assume that all code-words in $\mathcal{C}$ are equiprobable, and that we use nearest neighbour decoding (with respect to the Hamming distance on $\mathcal{V}$).

6.2 Examples of Codes

Here we will consider some simple examples of codes. They are easy to understand, but not very effective in terms of their rates or error-probabilities; we will consider more effective examples in later sections.

Example 6.3

The *repetition code* $\mathcal{R}_n$ over F consists of the words $\mathbf{u} = uu\ldots u \in \mathcal{V} = F^n$, where $u \in F$, so $M = |F| = q$. If F is a field then $\mathcal{R}_n$ is a linear code of dimension $k = 1$, spanned by the word (or vector) $11\ldots 1$. Fig. 6.1 shows the binary code $\mathcal{R}_3$ as a subset of $\mathcal{V} = F_2^3$, with the code-words represented by black vertices.

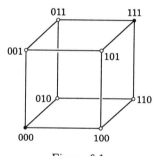

Figure 6.1

We saw in Chapter 5 that when $q = 2$ and n is odd, $\mathcal{R}_n$ corrects $(n-1)/2$ errors; by this we mean that if a code-word $\mathbf{u} \in \mathcal{R}_n$ is transmitted, and at most $(n-1)/2$ of its n symbols are transmitted incorrectly, then nearest neighbour decoding is always correct. A similar argument easily shows that for any q and n, if we use nearest neighbour decoding then $\mathcal{R}_n$ corrects $\lfloor (n-1)/2 \rfloor$ errors, where

$$\lfloor x \rfloor = \max\{m \in \mathbf{Z} \mid m \leq x\}$$

denotes the integer part of a real number x. This is excellent, but unfortunately (6.2) implies that $\mathcal{R}_n$ has rate $R = 1/n \to 0$ as $n \to \infty$, which is bad.

Example 6.4

The *parity-check code* $\mathcal{P}_n$ over a field $F = F_q$ consists of all vectors $\mathbf{u} = u_1 u_2 \ldots u_n \in V$ such that $\sum_i u_i = 0$; one can regard $u_1, \ldots u_{n-1}$ as information digits, and u_n as a check digit defined by $u_n = -u_1 - \cdots - u_{n-1}$. For instance, if $n = 3$ and $q = 2$ then $\mathcal{P}_3 = \{000, 011, 101, 110\}$, as shown in Fig. 6.2.

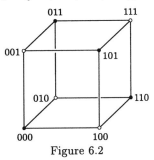

Figure 6.2

Since it is defined by a linear equation, $\mathcal{P}_n$ is a linear code. It has dimension $k = n - 1$, having basis $\mathbf{u}_1 = \mathbf{e}_1 - \mathbf{e}_n, \ldots, \mathbf{u}_{n-1} = \mathbf{e}_{n-1} - \mathbf{e}_n$ where the vectors $\mathbf{e}_i$ are the standard basis vectors $0 \ldots 010 \ldots 0$ of V. (To see this, note that each vector $\mathbf{u} = u_1 \ldots u_n \in \mathcal{P}_n$ can be written in a unique way as a linear combination $u_1 \mathbf{u}_1 + \cdots + u_{n-1} \mathbf{u}_{n-1}$ of the vectors $\mathbf{u}_i$.) Thus $M = q^{n-1}$ and $R = (n-1)/n$, so $R \to 1$ as $n \to \infty$, which is good.

Unfortunately, this code is almost useless for error-correcting, since it will detect a single error, but cannot correct it. Suppose that $\mathbf{u} = u_1 \ldots u_n \in \mathcal{P}_n$ is transmitted, and $\mathbf{v} = v_1 \ldots v_n \in V$ is received. The receiver computes $\sum_i v_i$ in F. If there is a single error, then exactly one digit v_i in $\mathbf{v}$ differs from the corresponding digit u_i in $\mathbf{u}$; since $\sum_i u_i = 0$ it follows that $\sum_i v_i \neq 0$, so the receiver knows that $\mathbf{v}$ is not a code-word and that an error must have occurred; however, there is no way of determining which digit is incorrect, since it is possible to obtain a code-word by changing *any* single digit of $\mathbf{v}$. Even worse, two or more compensating errors in $\mathbf{u}$ may go undetected.

Example 6.5

The *binary Hamming code* $\mathcal{H}_7$ is a linear code of length $n = 7$ over F_2. It was one of the first error-correcting codes discovered, having been introduced by the engineer Hamming in 1947 [Ha48, Ha50, Sh48] in frustration at the frequent crashes of the computer then being developed at Bell Laboratories (see [Th83] for a fascinating account of the early history of error-correcting codes).

To construct this code, we use Fig. 6.3, which shows a Venn diagram for three sets A, B and C. The regions corresponding to the sets $\overline{A} \cap \overline{B} \cap C, \overline{A} \cap B \cap \overline{C}, \overline{A} \cap B \cap C, A \cap \overline{B} \cap \overline{C}, A \cap \overline{B} \cap C, A \cap B \cap \overline{C}, A \cap B \cap C$ are numbered $1, 2, \ldots, 7$ in that order (we ignore $\overline{A} \cap \overline{B} \cap \overline{C}$); thus region number i is contained

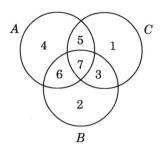

Figure 6.3

in A, B or C as the binary representation abc of the integer $i = 4a + 2b + c$ has a, b or c equal to 1. For instance, 5 is written as 101 in binary notation, so it corresponds to the region $A \cap \overline{B} \cap C$, while $\overline{A} \cap B \cap \overline{C}$ corresponds to 010, which represents 2.

We encode a block $\mathbf{a} = a_1 a_2 a_3 a_4$ of four binary information digits as a code-word $\mathbf{u} = u_1 \ldots u_7$ of length 7, by first defining $u_3 = a_1$, $u_5 = a_2$, $u_6 = a_3$, $u_7 = a_4$; we write these four digits in the regions numbered $3, 5, 6, 7$ respectively in Fig. 6.3. We define $u_4 = 0$ or 1, and write it in region 4, so that the binary sum of the four digits within the set A is 0, that is,

$$u_4 + u_5 + u_6 + u_7 = 0$$

in F_2. We define u_2 and u_1 similarly, using the sets B and C, so that

$$u_2 + u_3 + u_6 + u_7 = 0,$$
$$u_1 + u_3 + u_5 + u_7 = 0.$$

(Notice that the subscripts appearing in these three equations are those whose binary representations contain a 1 in the first, second and third positions a, b and c.) The code $\mathcal{H}_7$ consists of all the code-words $\mathbf{u} \in V = F_2^7$ formed in this way. Since $\mathcal{H}_7$ is defined by linear equations between the variables u_i, it is a linear code. There are $2^4 = 16$ choices for a_1, a_2, a_3 and a_4, and these determine $u_1, \ldots, u_7$ uniquely, so $M = |\mathcal{H}_7| = 16$; this also shows that $\mathcal{H}_7$ has dimension $k = 4$, with basis $\mathbf{u}_1 = 1110000$, $\mathbf{u}_2 = 1001100$, $\mathbf{u}_3 = 0101010$, $\mathbf{u}_4 = 1101001$ obtained by taking $a_1, a_2, a_3, a_4 = 1$ respectively, and the remaining terms $a_i = 0$.

This code corrects any single error in a code-word $\mathbf{u}$. Suppose that $\mathbf{u} \in \mathcal{H}_7$ is transmitted, and $\mathbf{v} \in V$ is received, where $\mathbf{v}$ differs from $\mathbf{u}$ only in its i-th digit v_i. The receiver computes

$$s_1 = v_4 + v_5 + v_6 + v_7,$$
$$s_2 = v_2 + v_3 + v_6 + v_7,$$
$$s_3 = v_1 + v_3 + v_5 + v_7$$

in F_2; these should all be 0, by the definition of $\mathcal{H}_7$, but the incorrect digit v_i causes at least one of them to be 1. Now v_i appears in the expression for s_j ($j = 1, 2, 3$) if and only if the j-th digit in the binary representation of i is 1, so $s_j = 0$ or 1 as this j-th digit is 0 or 1. This simply says that s_j is this j-th digit, so the word $\mathbf{s} = s_1 s_2 s_3$ is the binary representation of i. The receiver, having computed s_1, s_2 and s_3, can therefore locate the incorrect digit v_i, and can correct it by defining $u_i = v_i + 1$ in F_2.

As an illustration, suppose we want to encode $\mathbf{a} = 0110$. We define the information digits $u_3 = 0$, $u_5 = u_6 = 1$ and $u_7 = 0$, and then solve the three linear equations to obtain the check digits $u_4 = 0$, $u_2 = 1$ and $u_1 = 1$; the code-word transmitted is therefore $\mathbf{u} = 1100110$ ($= \mathbf{u}_2 + \mathbf{u}_3$). Now suppose that there is an error in the third digit, so that $\mathbf{v} = 1110110$ is received. The receiver computes $s_1 = 0 + 1 + 1 + 0 = 0$, $s_2 = 1 + 1 + 1 + 0 = 1$, $s_3 = 1 + 1 + 1 + 0 = 1$, and obtains $\mathbf{s} = s_1 s_2 s_3 = 011$, which is the binary representation of $i = 3$. The third digit of $\mathbf{v}$ is therefore changed to give $\Delta(\mathbf{v}) = 1100110$, which is correct, and the information digits $0, 1, 1, 0$ can be extracted from positions $3, 5, 6, 7$ of this word.

This code corrects any single error, but it fails if two or more errors occur. For instance, in the above example, suppose that there are errors in u_3 and u_4, so that $\mathbf{v}' = 1111110$ is received. The receiver computes $s_1' = 1$, $s_2' = 1$, $s_3' = 1$, giving $\mathbf{s}' = s_1' s_2' s_3' = 111$, which suggests an error in position $i = 7$, so $\mathbf{v}'$ is decoded as $\Delta(\mathbf{v}') = 1111111$, which is incorrect.

Exercise 6.2

Find the code-word in $\mathcal{H}_7$ representing the information digits 1101, and show how an error in its 6th symbol is corrected. What happens if there are errors in the 4th and 6th symbols?

Although the binary codes $\mathcal{R}_3$ and $\mathcal{H}_7$ both correct a single error, the rate $R = 4/7$ of $\mathcal{H}_7$ is significantly better than the rate $1/3$ of $\mathcal{R}_3$. In Chapter 7 we will generalise the construction of $\mathcal{H}_7$ to give a sequence of binary 1-error-correcting codes $\mathcal{H}_n$ ($n = 2^c - 1$), with rate $R \to 1$ as $n \to \infty$. You might like to think in advance how this could be done, replacing the three sets A, B, C with sets $A_1, \ldots, A_c$.

Example 6.6

Suppose that $\mathcal{C}$ is a code of length n over a field F. Then we can form a code of length $n + 1$ over F, called the *extended code* $\overline{\mathcal{C}}$, by adjoining an extra digit u_{n+1} to every code-word $\mathbf{u} = u_1 \ldots u_n \in \mathcal{C}$, chosen so that $u_1 + \cdots + u_{n+1} = 0$. Clearly $|\overline{\mathcal{C}}| = |\mathcal{C}|$, and if $\mathcal{C}$ is linear then so is $\overline{\mathcal{C}}$, with the same dimension. For instance, if $\mathcal{C} = V = F^n$ then $\overline{\mathcal{C}} = \mathcal{P}_{n+1} \subset F^{n+1}$.

Example 6.7

If $\mathcal{C}$ is a code of length n, we can form a *punctured code* $\mathcal{C}^\circ$ of length $n - 1$ by choosing a coordinate position i, and deleting the symbol u_i from each code-word $u_1 \ldots u_n \in \mathcal{C}$. In general, the structure of $\mathcal{C}^\circ$ depends on the choice of i.

6.3 Minimum Distance

Using nearest neighbour decoding, we will keep $\mathrm{Pr_E}$ low if we use code-words **u** which are far apart from each other, since then the transmitted code-word **u** is more likely to be the nearest code-word $\Delta(\mathbf{v})$ to the received word $\mathbf{v} \in \mathcal{V}$. We therefore define the *minimum distance* of a code $\mathcal{C}$ to be

$$d = d(\mathcal{C}) = \min\{d(\mathbf{u}, \mathbf{u}') \mid \mathbf{u}, \mathbf{u}' \in \mathcal{C}, \ \mathbf{u} \neq \mathbf{u}'\}, \tag{6.3}$$

the least Hamming distance between any two distinct code-words. A code of length n, with M code-words, and with minimum distance d is sometimes referred to as an (n, M, d)-code; if it is linear, of dimension k, it is called an $[n, k, d]$-code.

Our aim is to choose codes $\mathcal{C}$ for which d is large, so that $\mathrm{Pr_E}$ will be small. If $\mathcal{C}$ has M code-words, then finding d by means of (6.3) requires us to calculate and compare $\binom{M}{2} = M(M-1)/2$ distances, which could be quite tedious. However, this task is much simpler if $\mathcal{C}$ is linear, as we shall now show.

First we define the *weight* of any vector $\mathbf{v} = v_1 v_2 \ldots v_n \in \mathcal{V}$ to be

$$\mathrm{wt}(\mathbf{v}) = d(\mathbf{v}, \mathbf{0}), \tag{6.4}$$

where $\mathbf{0} = 00 \ldots 0 \in \mathcal{V}$. In other words, $\mathrm{wt}(\mathbf{v})$ is simply the number of subscripts i such that $v_i \neq 0$. It is easy to see that

$$d(\mathbf{u}, \mathbf{u}') = \mathrm{wt}(\mathbf{u} - \mathbf{u}')$$

for all $\mathbf{u}, \mathbf{u}' \in \mathcal{V}$.

Lemma 6.8

If $\mathcal{C}$ is a linear code, then its minimum distance d is given by

$$d = \min\{\mathrm{wt}(\mathbf{v}) \mid \mathbf{v} \in \mathcal{C}, \mathbf{v} \neq \mathbf{0}\}.$$

Proof

We have $d(\mathbf{u}, \mathbf{u}') = \mathrm{wt}(\mathbf{v})$ where $\mathbf{v} = \mathbf{u} - \mathbf{u}'$. Now $\mathcal{C}$ is a linear subspace of $\mathcal{V}$, so as **u** and **u'** range over all distinct pairs in $\mathcal{C}$, their difference $\mathbf{v} = \mathbf{u} - \mathbf{u}'$ ranges over all non-zero elements of C. It follows that $d(C)$, which is the minimum distance between such pairs $\mathbf{u}, \mathbf{u}'$, is equal to the minimum of these weights $\mathrm{wt}(\mathbf{v})$. $\qquad\qquad\qquad\qquad\qquad\qquad\qquad\qquad\qquad\qquad\qquad\qquad$ □

The advantage of this result is that it requires us to calculate and compare only $M - 1$ numbers, rather than $M(M - 1)/2$ in the case of non-linear codes. We will see in §7.3 that there are even better ways of calculating the minimum distance of a linear code.

Exercise 6.3

List all the codewords in the binary Hamming code $\mathcal{H}_7$ (Example 6.5), and use Lemma 6.8 to verify that the minimum distance is 3.

Exercise 6.4

Show that if C is a binary linear code of minimum distance d, then the extended code $\overline{C}$ has minimum distance d or $d + 1$ as d is even or odd. List the elements of the extended binary Hamming code $\overline{\mathcal{H}_7}$, and find its minimum distance.

We now consider how the minimum distance of a code affects its ability to correct errors. We say that a code C *corrects t errors*, or is *t-error-correcting*, if, whenever a code-word $\mathbf{u} \in C$ is transmitted and is then received with errors in at most t of its symbols, the resulting received word $\mathbf{v}$ is decoded correctly as $\mathbf{u}$; equivalently, whenever $\mathbf{u} \in C$ and $\mathbf{v} \in V$ satisfy $d(\mathbf{u}, \mathbf{v}) \leq t$, the decision rule Δ gives $\Delta(\mathbf{v}) = \mathbf{u}$.

Example 6.9

A repetition code $\mathcal{R}_3$ (over any alphabet) corrects one error, but not two (see §5.2 for the case $q = 2$). For instance, if $\mathbf{u} = 111$ is transmitted and $\mathbf{v} = 101$ is received (so there is one error), then nearest-neighbour decoding gives $\Delta(\mathbf{v}) = 111 = \mathbf{u}$. However, if $\mathbf{v} = 001$ is received (so there are two errors), then $\Delta(\mathbf{v}) = 000 \neq \mathbf{u}$.

If $\mathbf{u}$ is sent and $\mathbf{v}$ is received, we call the vector $\mathbf{e} = \mathbf{v} - \mathbf{u}$ the *error-pattern*, since its non-zero entries indicate where the errors in transmission have occurred, and what they are. The equation $\mathbf{v} = \mathbf{u} + \mathbf{e}$ shows that $\mathbf{v}$ consists of the transmitted word $\mathbf{u}$ plus the errors, represented by $\mathbf{e}$. The number of incorrect symbols is $d(\mathbf{u}, \mathbf{v}) = \mathrm{wt}(\mathbf{e})$, so a code corrects t errors if and only if it can correct all error-patterns $\mathbf{e} \in V$ of weight $\mathrm{wt}(\mathbf{e}) \leq t$.

Theorem 6.10

A code $\mathcal{C}$ of minimum distance d corrects t errors if and only if $d \geq 2t + 1$. (Equivalently, $\mathcal{C}$ corrects up to $\lfloor \frac{d-1}{2} \rfloor$ errors.)

Proof

($\Leftarrow$) Let $\mathcal{C}$ have minimum distance $d \geq 2t + 1$. Suppose that $\mathbf{u} \in \mathcal{C}$ is sent and $\mathbf{v} = \mathbf{u} + \mathbf{e} \in V$ is received, where the error-pattern $\mathbf{e} = \mathbf{v} - \mathbf{u}$ has weight $\mathrm{wt}(\mathbf{e}) \leq t$, so $d(\mathbf{u}, \mathbf{v}) \leq t$. For all $\mathbf{u}' \neq \mathbf{u}$ in $\mathcal{C}$ we have

$$d(\mathbf{u}, \mathbf{u}') \geq d \geq 2t + 1.$$

Now the triangle inequality (Lemma 5.8(c)) gives

$$d(\mathbf{u}, \mathbf{u}') \leq d(\mathbf{u}, \mathbf{v}) + d(\mathbf{v}, \mathbf{u}'),$$

so

$$d(\mathbf{v}, \mathbf{u}') \geq d(\mathbf{u}, \mathbf{u}') - d(\mathbf{u}, \mathbf{v}) \geq (2t + 1) - t = t + 1 > d(\mathbf{u}, \mathbf{v}).$$

Thus $\Delta(\mathbf{v}) = \mathbf{u}$, so decoding is correct, and $\mathcal{C}$ corrects t errors.

($\Rightarrow$) Suppose that $\mathcal{C}$ has minimum distance $d < 2t + 1$, so $d \leq 2t$. We can choose $\mathbf{u}, \mathbf{u}' \in \mathcal{C}$ so that $d(\mathbf{u}, \mathbf{u}') = d$. Then there exists a vector $\mathbf{v} \in V$ with

$$d(\mathbf{u}, \mathbf{v}) \leq t \quad \text{and} \quad d(\mathbf{u}', \mathbf{v}) \leq t.$$

(For instance, $\mathbf{u}$ and $\mathbf{u}'$ differ in exactly d symbols, and by changing $\lfloor d/2 \rfloor$ of those symbols u_i of $\mathbf{u}$ into the corresponding symbols u'_i of $\mathbf{u}'$ we get such a vector $\mathbf{v}$.) Now $\Delta(\mathbf{v})$ cannot be both $\mathbf{u}$ and $\mathbf{u}'$, so at least one of these two code-words, when transmitted and received as $\mathbf{v}$, is decoded incorrectly. Thus $\mathcal{C}$ does not correct t errors. $\qquad\square$

Example 6.11

A repetition code $\mathcal{R}_n$ of length n has minimum distance $d = n$, since $d(\mathbf{u}, \mathbf{u}') = n$ for all $\mathbf{u} \neq \mathbf{u}'$ in $\mathcal{R}_n$. This code therefore corrects $t = \lfloor \frac{n-1}{2} \rfloor$ errors.

Example 6.12

Exercise 6.3 shows that the Hamming code $\mathcal{H}_7$ has minimum distance $d = 3$, so it has $t = 1$ (as shown in §6.2). Similarly, $\overline{\mathcal{H}_7}$ has $d = 4$ (by Exercise 6.4), so this code also has $t = 1$.

Example 6.13

A parity-check code $\mathcal{P}_n$ of length n has minimum distance $d = 2$; for instance, the code-words $\mathbf{u} = 1(-1)0 \ldots 0$ and $\mathbf{u}' = \mathbf{0} = 00 \ldots 0$ are distance 2 apart, but no pair are distance 1 apart. It follows that the number of errors corrected by $\mathcal{P}_n$ is $t = \lfloor \frac{d-1}{2} \rfloor = 0$: for instance, $\mathbf{v} = 10 \ldots 0$ could be decoded as either $\mathbf{u}$ or $\mathbf{u}'$ (among others), each of which can give rise to $\mathbf{v}$ with a single error.

Although $\mathcal{P}_n$ is no use for correcting an error, it does at least *detect* one. More generally, suppose that a code $\mathcal{C}$ has minimum distance d, that a code-word $\mathbf{u} \in \mathcal{C}$ is sent, and that $\mathbf{v} = \mathbf{u} + \mathbf{e}$ is received, where $1 \leq \mathrm{wt}(\mathbf{e}) \leq d - 1$; then $\mathbf{v}$ cannot be a code-word, since $0 < d(\mathbf{u}, \mathbf{v}) < d$, so the receiver knows that there is at least one error among the symbols in $\mathbf{v}$. If $\mathrm{wt}(\mathbf{e}) = d$, however, it is possible that $\mathbf{v}$ is a code-word, in which case the receiver does not know whether $\mathbf{v}$ represents a correctly transmitted $\mathbf{v}$ or an incorrectly transmitted $\mathbf{u}$ (or even some other code-word). We therefore say that $\mathcal{C}$ *detects* $d - 1$ *errors*.

Example 6.14

The codes $\mathcal{R}_n$ and $\mathcal{P}_n$ have $d = n$ and 2 respectively, so $\mathcal{R}_n$ detects $n - 1$ errors, while $\mathcal{P}_n$ detects one; $\mathcal{H}_7$ has $d = 3$, so it detects two errors.

6.4 Hamming's Sphere-packing Bound

We have seen that a code $\mathcal{C}$ with minimum distance d corrects $t = \lfloor \frac{d-1}{2} \rfloor$ errors. The "spheres"[1]

$$S_t(\mathbf{u}) = \{ \mathbf{v} \in \mathcal{V} \mid d(\mathbf{u}, \mathbf{v}) \leq t \} \qquad (\mathbf{u} \in \mathcal{C}) \qquad (6.5)$$

are mutually disjoint, and each $S_t(\mathbf{u})$ consists entirely of vectors $\mathbf{v}$ decoded as $\mathbf{u}$ (though it need not contain all such $\mathbf{v}$). For good error-correction, we want the common radius t of these spheres to be large. However, to attain a good transmission-rate

$$R = \frac{\log_q M}{n}$$

we want the number M of these spheres to be large. If q and n are fixed, then since the spheres are disjoint, these two aims are in conflict with each other: we can think of $\mathcal{V}$ as an n-dimensional "box", of fixed size $q \times q \times \cdots \times q$, into which we are trying to pack a large number of non-intersecting large spheres. Clearly there is a limit to how far we can go in achieving this, and the next result, *Hamming's sphere-packing bound* [Ha50], makes this limit precise.

---

[1] Strictly speaking, these are balls, or solid spheres, being defined by $d(\mathbf{u}, \mathbf{v}) \leq t$ rather than $d(\mathbf{u}, \mathbf{v}) = t$, but we follow the convention in Coding Theory of calling them spheres.

Theorem 6.15

Let C be a q-ary t-error-correcting code of length n, with M code-words. Then

$$M\left(1 + \binom{n}{1}(q-1) + \binom{n}{2}(q-1)^2 + \cdots + \binom{n}{t}(q-1)^t\right) \leq q^n.$$

Proof

There are M spheres $S_t(\mathbf{u}) \subseteq \mathcal{V}$, one for each code-word $\mathbf{u} \in C$. As in Exercise 5.4, for each $\mathbf{u} \in C$ and for each i, the number of vectors $\mathbf{v} \in \mathcal{V}$ with $d(\mathbf{u}, \mathbf{v}) = i$ is $\binom{n}{i}(q-1)^i$: such a vector $\mathbf{v}$ must differ from $\mathbf{u}$ in exactly i of its n coordinate positions; these can be chosen in $\binom{n}{i}$ ways, and for each choice, there are $q-1$ ways of choosing each of these i coordinates of $\mathbf{v}$ to be different from the corresponding coordinate of $\mathbf{u}$. Summing this number for $i = 0, 1, \ldots, t$, we see from (6.5) that

$$|S_t(\mathbf{u})| = 1 + \binom{n}{1}(q-1) + \binom{n}{2}(q-1)^2 + \cdots + \binom{n}{t}(q-1)^t \qquad (6.6)$$

for each $\mathbf{u} \in C$. Now these M spheres are disjoint since $2t < d$, and they are all contained in a set $\mathcal{V}$ with q^n elements, so $M|S_t(\mathbf{u})| \leq q^n$, giving the required result. $\qquad \square$

Example 6.16

If we take $q = 2$ and $t = 1$ then Theorem 6.15 gives $M \leq 2^n/(1+n)$, so $M \leq \lfloor 2^n/(1+n) \rfloor$ since M must be an integer. Thus $M \leq 1, 1, 2, 3, 5, 9, 16, \ldots$ for $n = 1, 2, 3, 4, 5, 6, 7, \ldots$.

Corollary 6.17

Every t-error-correcting linear $[n, k]$-code C over F_q satisfies

$$\sum_{i=0}^{t} \binom{n}{i}(q-1)^i \leq q^{n-k}.$$

Proof

Since $\dim(C) = k$ we have $M = q^k$; now divide by q^k in Theorem 6.15. $\qquad \square$

In a linear $[n, k]$-code C, each code-word has n digits, k of which can be regarded as carrying information, while the remaining $n - k$ are check digits. Corollary 6.17 therefore gives us a lower bound

$$n - k \geq \log_q\left(\sum_{i=0}^{t} \binom{n}{i}(q-1)^i\right)$$

on the number of check digits required to correct t errors.

A code $\mathcal{C}$ is *perfect* if it attains equality in Theorem 6.15 (equivalently in Corollary 6.17, in the case of a linear code). This is equivalent to requiring that the disjoint spheres $S_t(\mathbf{u})$ ($\mathbf{u} \in \mathcal{C}$) fill $\mathcal{V}$ completely, so that every $\mathbf{v} \in \mathcal{V}$ is within distance at most t of exactly one code-word $\mathbf{u}$. (Such a perfect sphere-packing is impossible in a euclidean space $\mathbf{R}^n$ of dimension $n > 1$, since there are always unfilled gaps between the spheres; the best possible packing in the plane is well-known — and obvious — but the corresponding problem in $\mathbf{R}^3$ was not solved until 1998, by Thomas Hales: see www.math.lsa.umich.edu/~hales. See [CS92] and [Th83] for connections between euclidean sphere-packing and coding theory.)

Exercise 6.5

Show that a code is perfect if and only if, for some t, nearest-neighbour decoding corrects all error-patterns of weight at most t, and none of weight greater than t.

Example 6.18

Let $\mathcal{C}$ be a binary repetition code $\mathcal{R}_n$ of odd length n. This is a linear code with $k = 1, q = 2$ and $t = \lfloor \frac{n-1}{2} \rfloor = \frac{n-1}{2}$, so in Corollary 6.17 we have $n - k = n - 1$. Now $q - 1 = 1$, and $\binom{n}{i} = \binom{n}{n-i}$ for all i, so

$$\sum_{i=0}^{t} \binom{n}{i}(q-1)^i = \sum_{i=0}^{t} \binom{n}{i} = \frac{1}{2}\sum_{i=0}^{n} \binom{n}{i} = \frac{1}{2}.2^n = 2^{n-1}.$$

Thus the bound in Corollary 6.17 is attained, so this code is perfect. However, if n is even or if $q > 2$ then $\mathcal{R}_n$ is not perfect. Fig. 6.4 illustrates why the binary code $\mathcal{R}_3$ is perfect, by showing how the eight elements of $\mathcal{V} = F_2^3$ are partitioned into two sets $S_1(\mathbf{u})$, coloured black and white as $\mathbf{u} = 000$ or 111; there is a similar partition of F_2^n into two sets for all odd n.

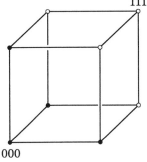

Figure 6.4

Example 6.19

The binary Hamming code $\mathcal{H}_7$ is a linear $[7, 4]$-code, that is, $n = 7$ and $k = 4$. It has $q = 2$ and $t = 1$, so

$$\sum_{i=0}^{t} \binom{n}{i} (q - 1)^i = 1 + \binom{7}{1} = 8 = q^{n-k}$$

and the code is perfect. We will see in Chapter 7 that this is one of a family of binary Hamming codes $\mathcal{H}_n$ ($n = 2^c - 1$), all of which are perfect.

Exercise 6.6

The binary Hamming code $\mathcal{H}_7$ is used, where the information channel Γ is a BSC with $P > \frac{1}{2}$, and Δ is nearest neighbour decoding; find the error probability $\Pr_E$. Show that $\Pr_E \approx 21Q^2$ if $Q = \overline{P}$ is small.

Exercise 6.7

Let $\mathcal{C}$ be the extended binary Hamming code $\overline{\mathcal{H}_7}$ (see Exercise 6.4). Find how many vectors $\mathbf{v} \in V = F_2^8$ are covered by the spheres $S_t(\mathbf{u})$, where $\mathbf{u} \in \mathcal{C}$, and hence show that this code is not perfect.

If $\mathcal{C}$ is any binary code then Theorem 6.15 gives

$$2^n \geq M \binom{n}{t} = 2^{nR} \binom{n}{t}.$$

Thus $2^{n(1-R)} \geq \binom{n}{t}$, so taking logarithms gives

$$1 - R \geq \frac{1}{n} \log_2 \binom{n}{t}.$$

If we apply Stirling's approximation $n! \sim (n/e)^n \sqrt{2\pi n}$ (see [Fi83] or [La83], for instance) to the three factorials in $\binom{n}{t} = n!/t!\,(n-t)!$, we find that the right-hand side approaches $H_2(t/n)$ as $n \to \infty$ with t/n constant, where H_2 is the binary entropy function, defined in §3.1 (see Exercise 6.8). In the limit we get

$$H_2\left(\frac{t}{n}\right) \leq 1 - R, \tag{6.7}$$

which is *Hamming's upper bound* on the proportion t/n of errors corrected by binary codes of rate R, as $n \to \infty$. Fig. 6.5 shows the region allowed by this inequality; notice that we restrict to $t/n < 1/2$, since $d \leq n$ and Theorem 6.10 gives $t = \lfloor (d-1)/2 \rfloor$.

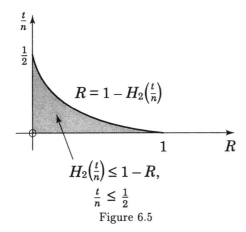

Figure 6.5

It is a useful exercise to determine which points in this region correspond to various binary codes, such as the repetition, parity-check and Hamming codes.

Exercise 6.8

Prove that $\frac{1}{n} \log_2 \binom{n}{t} \to H_2(t/n)$ as $n \to \infty$ with t/n constant.

6.5 The Gilbert–Varshamov Bound

In order to maximise the rate $R = \frac{1}{n} \log_q M$, while retaining good error-correcting properties, we are interested in finding codes with the largest possible value of $M = |\mathcal{C}|$, for given values of q, n and t (or equivalently d). Let $A_q(n, d)$ denote the greatest number of code-words in any q-ary code of length n and minimum distance d, where $d \leq n$. Hamming's sphere-packing bound (Theorem 6.15) gives an upper bound for $A_q(n, d)$ by showing that

$$A_q(n, d)\left(1 + \binom{n}{1}(q - 1) + \binom{n}{2}(q - 1)^2 + \cdots + \binom{n}{t}(q - 1)^t\right) \leq q^n,$$

where $t = \lfloor (d - 1)/2 \rfloor$ by Theorem 6.10.

Example 6.20

If $q = 2$ and $d = 3$ then $t = 1$, so as in Example 6.16 we find that $A_2(n, 3) \leq \lfloor 2^n/(n + 1) \rfloor$. Thus for $n = 3, 4, 5, 6, 7, \ldots$ we have $A_2(n, 3) \leq 2, 3, 5, 9, 16, \ldots$.

Exercise 6.9

Find upper bounds for $A_3(n, 3)$ corresponding to those given for $A_2(n, 3)$ in Example 6.20. What does Hamming's sphere-packing bound imply about $A_2(n, 4)$ and $A_2(n, 5)$?

A similar argument gives a *lower* bound for $A_q(n, d)$, showing that for given q, n and d, there exists a code with at least a given number of code-words. This is the *Gilbert–Varshamov bound* [Gi52, Va57]:

Theorem 6.21

If $q \geq 2$ and $n \geq d \geq 1$ then

$$A_q(n, d)\left(1 + \binom{n}{1}(q - 1) + \binom{n}{2}(q - 1)^2 + \cdots + \binom{n}{d-1}(q - 1)^{d-1}\right) \geq q^n.$$

Proof

Among all the codes with the chosen values of q, n and d, let C have the maximum number of code-words, so $M = |C| = A_q(n, d)$. The spheres

$$S_{d-1}(\mathbf{u}) = \{\mathbf{v} \in V \mid d(\mathbf{u}, \mathbf{v}) \leq d - 1\},$$

where $\mathbf{u} \in C$, must cover V: for if some $\mathbf{v} \in V$ is not in any $S_{d-1}(\mathbf{u})$, then $d(\mathbf{u}, \mathbf{v}) \geq d$ for all $\mathbf{u} \in C$, so the code $C' = C \cup \{\mathbf{v}\}$ has the same values of q, n and d, and has $|C'| > |C|$, contradicting the choice of C. By the argument used to prove (6.6), each of the M spheres $S_{d-1}(\mathbf{u})$ contains $\sum_{i=0}^{d-1} \binom{n}{i}(q - 1)^i$ vectors; between them, these spheres contain all the q^n vectors in V, so the result follows.   □

Example 6.22

If we take $q = 2$ and $d = 3$ again (so that $t = 1$), then Theorem 6.21 gives

$$A_2(n, 3)\left(1 + n + \frac{n(n - 1)}{2}\right) \geq 2^n$$

for all $n \geq 3$, so $A_2(n, 3) \geq 2^{n+1}/(n^2 + n + 2)$. Since $A_q(n, d)$ is an integer, we therefore have

$$A_2(n, 3) \geq \lceil 2^{n+1}/(n^2 + n + 2)\rceil.$$

For $n = 3, 4, 5, 6, 7, \ldots$ this gives $A_2(n, 3) \geq 2, 2, 2, 3, 5, \ldots$. If we compare these lower bounds with the upper bounds given in Example 6.20, we see that $A_2(3, 3) = 2$: for example, the binary repetition code R_3 attains this bound. When $n = 4$ we have $2 \leq A_2(4, 3) \leq 3$, so $A_2(4, 3) = 2$ or 3.

Exercise 6.10

Show that $A_2(4,3) = 2$, and find a code attaining this bound.

Exercise 6.11

Find a lower bound for $A_3(n,3)$.

For most values of q, n and d, there is a significant gap between our upper and lower bounds for $A_q(n,d)$, and it can be a difficult problem to determine its exact value; indeed, in many cases it is still unknown. In certain cases, the existence of a specific code tells us its value: thus for $q = 2, d = 3$ and $n = 7$ the Hamming code $\mathcal{H}_7$ attains the upper bound $M \leq 16$ implied by Theorem 6.15, so $A_2(7,3) = 16$. More generally, we will see in §7.4 that if n has the form $2^c - 1$ then $A_2(n,3)$ attains the upper bound 2^{n-c}.

In the binary case, Theorem 6.21 takes the form

$$A_2(n,d)\left(1 + \binom{n}{1} + \binom{n}{2} + \cdots + \binom{n}{d-1}\right) \geq 2^n.$$

Now Exercise 5.7 gives

$$\sum_{i \leq nQ} \binom{n}{i} \leq 2^{nH_2(Q)}$$

for $Q < \frac{1}{2}$, so

$$\log_2 A_2(n,d) \geq n\left(1 - H_2\left(\frac{d-1}{n}\right)\right)$$

for $d \leq \lfloor n/2 \rfloor$. Since a binary code has rate $R = \frac{1}{n}\log_2 M$, this proves that if $d \leq \lfloor n/2 \rfloor$ there exists such a code with length n, minimum distance d, and rate

$$R \geq 1 - H_2\left(\frac{d-1}{n}\right).$$

We can compare this lower bound with Hamming's asymptotic upper bound

$$R \leq 1 - H_2\left(\frac{t}{n}\right),$$

proved in §6.4, where $t = \lfloor (d-1)/2 \rfloor$ by Theorem 6.10. Fig. 6.6 shows the region defined by these two bounds on R.

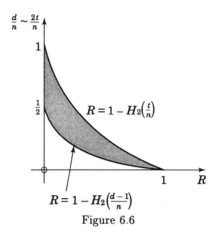

Figure 6.6

6.6 Hadamard Matrices and Codes

Many mathematical structures can be used to provide codes. An interesting class of codes can be constructed from a class of matrices called the Hadamard matrices. First we examine the elementary properties of these matrices (see [Ha67, MS77] for further details).

Hadamard was interested in how large the determinant of a real $n \times n$ matrix $H = (h_{ij})$ could be, for each given value of n. For this to make sense, one has to bound the entries of H, and there is no loss of generality in supposing that $|h_{ij}| \leq 1$ for all i and j. Under these conditions, Hadamard proved that $|\det H| \leq n^{n/2}$, with equality if and only if

(a) each $h_{ij} = \pm 1$, and

(b) distinct rows $\mathbf{r}_i$ of H are orthogonal, that is, $\mathbf{r}_i.\mathbf{r}_j = 0$ for all $i \neq j$.

An $n \times n$ matrix H satisfying (a) and (b) is called a *Hadamard matrix* of order n. Since (a) implies that $\mathbf{r}_i.\mathbf{r}_i = n$ for all i, it follows that HH^{T} is the diagonal matrix

$$HH^{\mathrm{T}} = \begin{pmatrix} n & 0 & \cdots & 0 \\ 0 & n & \cdots & 0 \\ \vdots & \vdots & \ddots & \vdots \\ 0 & 0 & \cdots & n \end{pmatrix} = nI_n ; \tag{6.8}$$

here H^{T} denotes the transpose of H, and I_n is the $n \times n$ identity matrix. Since $\det H^{\mathrm{T}} = \det H$, it follows from (6.8) that

$$(\det H)^2 = \det(nI_n) = n^n,$$

so $|\det H| = n^{n/2}$. Thus all Hadamard matrices attain Hadamard's upper bound; we will omit the converse, which is harder and is not needed here.

For aesthetic and typographical reasons, we will follow the convention of indicating an entry -1 in a Hadamard matrix by simply $-$.

Example 6.23

The matrices $H = (1)$ and $\left(\begin{smallmatrix} 1 & 1 \\ 1 & - \end{smallmatrix}\right)$ are Hadamard matrices of order 1 and 2, with $|\det H| = 1$ and 2 respectively.

Exercise 6.12

Find all the Hadamard matrices of orders 1 and 2.

The following simple result enables us to construct large Hadamard matrices from smaller ones.

Lemma 6.24

Let H be a Hadamard matrix of order n, and let

$$H' = \begin{pmatrix} H & H \\ H & -H \end{pmatrix}.$$

Then H' is a Hadamard matrix of order $2n$.

Exercise 6.13

Prove Lemma 6.24.

Corollary 6.25

There is a Hadamard matrix of order 2^m for each integer $m \geq 0$.

Proof

Start with $H = (1)$, and apply Lemma 6.24 m times.   □

Example 6.26

The Hadamard matrices of order 2^m obtained by this method are called *Sylvester matrices*. For instance, taking $m = 1$ gives $\left(\begin{smallmatrix} 1 & 1 \\ 1 & - \end{smallmatrix}\right)$, and for $m = 2$ we get

$$\begin{pmatrix} 1 & 1 & 1 & 1 \\ 1 & - & 1 & - \\ 1 & 1 & - & - \\ 1 & - & - & 1 \end{pmatrix}.$$

However, Hadamard matrices do not exist for all orders. For example, there are none of odd order $n > 1$:

Lemma 6.27

If there is a Hadamard matrix H of order $n > 1$, then n is even.

Proof

The orthogonality of distinct rows $\mathbf{r}_i$ and $\mathbf{r}_j$ gives $h_{i1}h_{j1} + \cdots + h_{in}h_{jn} = 0$. Each $h_{ik}h_{jk} = \pm 1$, so n must be even. $\square$

By working a little harder, we obtain the following stronger restriction on the order of a Hadamard matrix:

Lemma 6.28

If there is a Hadamard matrix H of order $n > 2$, then n is divisible by 4.

Proof

Multiplying any column of H by -1 preserves the Hadamard property, so we may assume that the entries in the first row $\mathbf{r}_1$ are all 1. Each row $\mathbf{r}_i$ ($i \neq 1$) is orthogonal to $\mathbf{r}_1$, so it has $n/2$ terms equal to 1, and $n/2$ equal to -1. Permuting the columns (which also preserves the Hadamard property), we may assume that

$$\mathbf{r}_2 = \begin{pmatrix} 1 & 1 & \cdots & 1 & -1 & -1 & \cdots & -1 \end{pmatrix}.$$

Now suppose that the first and last $n/2$ entries of $\mathbf{r}_3$ contain u and v terms equal to 1 respectively (the other terms being -1). Then

$$0 = \mathbf{r}_1.\mathbf{r}_3 = u - \left(\frac{n}{2} - u\right) + v - \left(\frac{n}{2} - v\right) = 2u + 2v - n$$

and

$$0 = \mathbf{r}_2.\mathbf{r}_3 = u - \left(\frac{n}{2} - u\right) - v + \left(\frac{n}{2} - v\right) = 2u - 2v,$$

so $u = v$, and hence $n = 2u + 2v = 4u$ is divisible by 4. $\square$

It is conjectured that the converse is true, that there is a Hadamard matrix of order n for each n divisible by 4. This is still an open problem. The relevance of Hadamard matrices to Coding Theory rests on the following result.

Theorem 6.29

Each Hadamard matrix H of order n gives rise to a binary code of length n, with $M = 2n$ code-words and minimum distance $d = n/2$.

Proof

First we form $2n$ vectors $\pm\mathbf{r}_1, \ldots, \pm\mathbf{r}_n \in \mathbf{R}^n$ from the rows $\mathbf{r}_i$ of H. The orthogonality of the rows implies that these vectors are all distinct. By changing each entry -1 into 0 we get $2n$ vectors with entries $0, 1$; we can regard these as elements of $V = F_2^n$, so they form a binary code C. By their construction, these code-words have the form $\mathbf{u}_1, \bar{\mathbf{u}}_1, \ldots, \mathbf{u}_n, \bar{\mathbf{u}}_n$, where $\bar{\mathbf{u}} = \mathbf{1} - \mathbf{u}$. We have $d(\mathbf{u}_i, \bar{\mathbf{u}}_i) = n$ for all i, since $\mathbf{u}_i$ and $\bar{\mathbf{u}}_i$ differ in all n coordinates, and condition (b) easily implies that all other pairs of distinct code-words are distance $n/2$ apart, so C has minimum distance $d = n/2$.   □

Exercise 6.14

Find all the code-words obtained in the above way from the Hadamard matrix H in Example 6.26. Do they form a linear code?

Any code C constructed as in Theorem 6.29 is called a *Hadamard code* of length n. Such a code, of length 32, was used to transmit pictures from the 1969 Mariner space-probe.

Exercise 6.15

Construct a Hadamard matrix of order 8, and hence a Hadamard code of length 8. What is its rate? How many errors does it correct, and how many does it detect?

If n is not a power of 2 then neither is $2n$, so a Hadamard code of such a length n cannot be linear. The transmission rate of any Hadamard code of length n is
$$R = \frac{\log_2(2n)}{n} = \frac{1 + \log_2 n}{n} \to 0 \quad \text{as} \quad n \to \infty.$$
The number of errors corrected (if $n > 2$) is
$$t = \left\lfloor \frac{d-1}{2} \right\rfloor = \left\lfloor \frac{n-2}{4} \right\rfloor = \frac{n}{4} - 1,$$
by Theorems 6.10 and 6.29 and Lemma 6.28, so the proportion of errors corrected is
$$\frac{t}{n} = \frac{1}{4} - \frac{1}{n} \to \frac{1}{4} \quad \text{as} \quad n \to \infty.$$

6.7 Supplementary Exercises

Exercise 6.16

Find a cubic polynomial $f(x)$ which is irreducible over $\mathbf{Z}_2$, and use it to construct a field F_8 of order $q = 8$. Show that there are precisely two such polynomials, and that the corresponding fields are isomorphic.

Exercise 6.17

Show that for each prime $p \equiv 3 \bmod (4)$, the polynomial $f(x) = x^2 + 1$ is irreducible over $\mathbf{Z}_p$. Use this to construct a field F_q of order $q = p^2$. For which primes p does the polynomial $x^2 + x + 1$ give rise to a field of order $q = p^2$?

Exercise 6.18

Prove the *Singleton bound*: if a code over F_q has length n, minimum distance d, and M code-words, then $\log_q M \le n - d + 1$. (Hint: puncture the code $d - 1$ times.) Give some examples of *maximum distance separable* (MDS) codes, those which attain this bound.

Exercise 6.19

Calculate and factorise the numbers

$$1 + \binom{23}{1} + \binom{23}{2} + \binom{23}{3} \qquad \text{and} \qquad 1 + \binom{11}{1} \cdot 2 + \binom{11}{2} \cdot 2^2.$$

What do these results suggest about the possible existence of perfect codes?

Exercise 6.20

Show that if C_i is an (n, M_i, d_i)-code in $V = F_q^n$ for $i = 1, 2$, then

$$C_1 \oplus C_2 = \{ (\mathbf{x}, \mathbf{y}) \in V \oplus V \mid \mathbf{x} \in C_1, \, \mathbf{y} \in C_2 \}$$

is a $(2n, M_1 M_2, d)$-code, where $d = \min(d_1, d_2)$. Show also that

$$C_1 * C_2 = \{ (\mathbf{x}, \mathbf{x} + \mathbf{y}) \in V \oplus V \mid \mathbf{x} \in C_1, \, \mathbf{y} \in C_2 \}$$

is a $(2n, M_1 M_2, d')$-code, where $d' = \min(2d_1, d_2)$. Show that if each C_i is linear, of dimension k_i, then $C_1 \oplus C_2$ and $C_1 * C_2$ are both linear, of dimension $k_1 + k_2$.

Exercise 6.21

If we ignore the hyphens, then an International Standard Book Number (ISBN) is a code-word $w = a_1 \ldots a_{10}$ of length 10 over $\mathbf{Z}_{11} = \{0, 1, \ldots, 9, X\}$, with X denoting "ten". The digits $a_1, \ldots, a_9$ are information digits, indicating country, publisher, etc., while a_{10} is a check-digit defined by $a_1 + 2a_2 + \cdots + 10a_{10} \equiv 0 \bmod (11)$. Show that this code can detect any single incorrect digit, and also the transposition of two digits (the two most common human errors). Which of the following are valid ISBNs ?

$$3\text{-}540\text{-}76197\text{-}7, \quad 3\text{-}540\text{-}76179\text{-}7, \quad 3\text{-}541\text{-}76197\text{-}7.$$

7
Linear Codes

Report me and my cause aright. (*Hamlet*)

In Chapter 6 we considered several examples of linear codes, and in Lemma 6.8 we have already seen one advantage of dealing with them, namely that calculating the minimum distance of a linear code is easier than for general codes. In this chapter we will study linear codes in greater detail, noting other advantages to be obtained by applying elementary linear algebra and matrix theory, including an even simpler method for calculating the minimum distance. The theoretical background required includes such topics as linear independence, dimension, and row and column operations. These are normally covered in any first-year university linear algebra course; although such courses often restrict attention to vector spaces and matrices over the fields of real or complex numbers, all the important results and techniques we need extend in the obvious way to arbitrary fields, including finite fields. Throughout this chapter, we will assume that the alphabet F is the finite field F_q of order q, for some prime-power $q = p^e$.

7.1 Matrix Description of Linear Codes

One can specify a linear code $C \subseteq V = F^n$ by giving a basis $\mathbf{u}_1, \ldots, \mathbf{u}_k$ for C, so the code-words $\mathbf{u} \in C$ are the linear combinations $a_1 \mathbf{u}_1 + \cdots + a_k \mathbf{u}_k$ ($a_i \in F$)

of the basis vectors. This requires us to list just k vectors, where $k = \dim(C)$, rather than all $M = q^k$ vectors in C. A useful way of specifying a basis is to give a *generator matrix* G for C: this has k rows and n columns, each row being one of the basis vectors for C. (Note that C does not determine G uniquely: a subspace may have many bases, and the vectors in a given basis may be written in any order.)

Example 7.1

The repetition code $\mathcal{R}_n$ over F has a single basis vector $\mathbf{u}_1 = 11 \ldots 1$, so it has a generator matrix

$$G = (1 \quad 1 \quad \ldots \quad 1)$$

with one row and n columns.

Example 7.2

The parity-check code $\mathcal{P}_n$ over F has basis $\mathbf{u}_1, \ldots, \mathbf{u}_{n-1}$, where each $\mathbf{u}_i = \mathbf{e}_i - \mathbf{e}_n$ in terms of the standard basis vectors $\mathbf{e}_1, \ldots, \mathbf{e}_n$ of $\mathcal{V}$. It therefore has a generator matrix

$$G = \begin{pmatrix} 1 & & & -1 \\ & 1 & & -1 \\ & & \ddots & \vdots \\ & & & 1 & -1 \end{pmatrix}$$

with $n - 1$ rows and n columns, where the missing entries are all 0.

Example 7.3

A basis $\mathbf{u}_1 = 1110000$, $\mathbf{u}_2 = 1001100$, $\mathbf{u}_3 = 0101010$, $\mathbf{u}_4 = 1101001$ for the binary Hamming code $\mathcal{H}_7$ was given in Example 6.5. This code therefore has a generator matrix

$$G = \begin{pmatrix} 1 & 1 & 1 & 0 & 0 & 0 & 0 \\ 1 & 0 & 0 & 1 & 1 & 0 & 0 \\ 0 & 1 & 0 & 1 & 0 & 1 & 0 \\ 1 & 1 & 0 & 1 & 0 & 0 & 1 \end{pmatrix}.$$

If a linear code C has dimension k, then we can regard the k-dimensional vector space $\mathcal{A} = F^k$ as a source, encoded by the linear isomorphism $\mathcal{A} \to C \subseteq \mathcal{V} = F^n$ given by the matrix G. Specifically, each word $\mathbf{a} = a_1 \ldots a_k \in \mathcal{A}$ is encoded as the code-word $\mathbf{u} = \mathbf{a}G \in C$, giving an isomorphism $\mathbf{a} \mapsto \mathbf{u}$ between the vector spaces $\mathcal{A}$ and C. Thus encoding is multiplication by a fixed matrix, which is easy to perform.

Example 7.4

The repetition code $\mathcal{R}_n$ has $k = 1$, so $\mathcal{A} = F^1 = F$. Each $\mathbf{a} = a \in \mathcal{A}$ is encoded as $\mathbf{u} = \mathbf{a}G = a \ldots a \in \mathcal{R}_n$.

Example 7.5

If $\mathcal{C} = \mathcal{P}_n$ then $k = n - 1$, so $\mathcal{A} = F^{n-1}$. Each $\mathbf{a} = a_1 \ldots a_{n-1} \in \mathcal{A}$ is encoded as $\mathbf{u} = \mathbf{a}G = a_1 \ldots a_{n-1}a_n$, where $a_n = -(a_1 + \cdots + a_{n-1})$, so $\sum_i a_i = 0$.

Example 7.6

If $\mathcal{C} = \mathcal{H}_7$ then $n = 7$ and $k = 4$, so $\mathcal{A} = F_2^4$. Each $\mathbf{a} = a_1 \ldots a_4 \in \mathcal{A}$ is encoded as $\mathbf{u} = \mathbf{a}G \in \mathcal{H}_7 \subset F_2^7$. For instance, in Example 6.5 we encoded $\mathbf{a} = 0110$ as

$$\mathbf{u} = \mathbf{a}G = (0 \quad 1 \quad 1 \quad 0) \begin{pmatrix} 1 & 1 & 1 & 0 & 0 & 0 & 0 \\ 1 & 0 & 0 & 1 & 1 & 0 & 0 \\ 0 & 1 & 0 & 1 & 0 & 1 & 0 \\ 1 & 1 & 0 & 1 & 0 & 0 & 1 \end{pmatrix} = (1 \quad 1 \quad 0 \quad 0 \quad 1 \quad 1 \quad 0).$$

Given a generator matrix G for a linear code $\mathcal{C}$, it can be quite tedious to determine whether a vector $\mathbf{v} \in V$ is in $\mathcal{C}$, and if not, which element $\mathbf{u} \in \mathcal{C}$ is closest to it. To make this easier, we look for an alternative matrix description of $\mathcal{C}$. An effective way of doing this is to give a set of $n - k$ simultaneous linear equations which define the elements of $\mathcal{C}$, so that a vector $\mathbf{v} \in V$ lies in $\mathcal{C}$ if and only if its entries v_i satisfy these equations.

Example 7.7

The repetition code $\mathcal{R}_n$ consists of the vectors $\mathbf{v} = v_1 \ldots v_n \in V$ satisfying $v_1 = \cdots = v_n$, which can be regarded as a set of $n - k = n - 1$ simultaneous linear equations $v_i - v_n = 0$ $(i = 1, \ldots, n - 1)$.

Example 7.8

The parity-check code $\mathcal{P}_n$ (which has $n - k = 1$) is the subspace of V defined by the single linear equation $v_1 + \cdots + v_n = 0$.

Example 7.9

The Hamming code $\mathcal{H}_7$ consists of the vectors $\mathbf{v} \in V = F_2^7$ satisfying

$$v_4 + v_5 + v_6 + v_7 = 0,$$
$$v_2 + v_3 + v_6 + v_7 = 0,$$
$$v_1 + v_3 + v_5 + v_7 = 0.$$

In general, c independent linear equations define a subspace of $\mathcal{V}$ of dimension $n - c$, so we require $c = n - k$ independent equations to specify $\mathcal{C}$. These are called *parity-check equations*, and their matrix H of coefficients, which has n columns and $n - k$ independent rows, is called a *parity-check matrix* for $\mathcal{C}$. The linear equations can be written in the form $\mathbf{v}H^{\mathrm{T}} = \mathbf{0}$, where H^{T} denotes the transpose of the matrix H, so we have the following useful and efficient test for code-words of $\mathcal{C}$:

Lemma 7.10

Let $\mathcal{C}$ be a linear code, contained in $\mathcal{V}$, with parity-check matrix H, and let $\mathbf{v} \in \mathcal{V}$. Then $\mathbf{v} \in \mathcal{C}$ if and only if $\mathbf{v}H^{\mathrm{T}} = \mathbf{0}$.

Example 7.11

Using the equations $v_i - v_n = 0$ for $i = 1, \ldots, n - 1$ we see that the matrix

$$H = \begin{pmatrix} 1 & & & -1 \\ & 1 & & -1 \\ & & \ddots & \vdots \\ & & & 1 & -1 \end{pmatrix},$$

with $n - 1$ rows and n columns, is a parity-check matrix for $\mathcal{R}_n$.

Example 7.12

The equation $v_1 + \cdots + v_n = 0$ shows that we can take

$$H = (1 \quad 1 \quad \ldots \quad 1),$$

with one row and n columns, as a parity-check matrix for $\mathcal{P}_n$.

Example 7.13

The three linear equations given in Example 7.9 for $\mathcal{H}_7$ provide a parity-check matrix

$$H = \begin{pmatrix} 0 & 0 & 0 & 1 & 1 & 1 & 1 \\ 0 & 1 & 1 & 0 & 0 & 1 & 1 \\ 1 & 0 & 1 & 0 & 1 & 0 & 1 \end{pmatrix}.$$

Exercise 7.1

If $\mathcal{C}$ is a linear code, with generator matrix G and parity-check matrix H, find a generator matrix $\overline{G}$ and a parity-check matrix $\overline{H}$ for the extended code $\overline{\mathcal{C}}$.

Exercise 7.2

If $C_1, C_2 \subseteq V$ are linear codes, with generator matrices G_1, G_2 and parity-check matrices H_1, H_2, explain how to find a generator matrix for $C_1 + C_2$ and a parity-check matrix for $C_1 \cap C_2$.

One can view H as the matrix of a linear transformation $h : V \to W = F^{n-k}$, sending each $\mathbf{v} \in V$ to $h(\mathbf{v}) = \mathbf{v}H^{\mathrm{T}} \in W$, so Lemma 7.10 asserts that C is the kernel $\ker(h)$ of h, the set of vectors sent to $\mathbf{0}$. The image $\mathrm{im}(h)$ of h is the subspace of W spanned by the columns of H. The dimension theorem $\dim(V) = \dim(\ker(h)) + \dim(\mathrm{im}(h))$ implies that $\dim(\mathrm{im}(h)) = n - k$, so h maps V onto W. The $n - k$ rows of H (representing the linear equations defining C) are linearly independent, so they form the basis of a linear subspace $D \subseteq V$ of dimension $n - k$; this is a linear code, with generator matrix H, called the *dual code* of C.

The codes C and D are related by the concept of orthogonality. Just as in euclidean space $\mathbf{R}^n$, one can define a scalar product on $V = F^n$ by

$$\mathbf{u}.\mathbf{v} = u_1 v_1 + \cdots + u_n v_n \in F \tag{7.1}$$

where $\mathbf{u} = u_1 \ldots u_n$ and $\mathbf{v} = v_1 \ldots v_n$ are any vectors in V. This is linear in both variables, meaning that

$$(a\mathbf{u}_1 + b\mathbf{u}_2).\mathbf{v} = a(\mathbf{u}_1.\mathbf{v}) + b(\mathbf{u}_2.\mathbf{v}) \quad \text{and} \quad \mathbf{u}.(a\mathbf{v}_1 + b\mathbf{v}_2) = a(\mathbf{u}.\mathbf{v}_1) + b(\mathbf{u}.\mathbf{v}_2)$$

for all $a, b \in F$. We define $\mathbf{u}$ and $\mathbf{v}$ to be *orthogonal* if $\mathbf{u}.\mathbf{v} = 0$. Unlike in $\mathbf{R}^n$, a non-zero vector can be orthogonal to itself: for instance, if $\mathbf{u} = \mathbf{e}_1 + \mathbf{e}_2$ then $\mathbf{u}.\mathbf{u} = 1^2 + 1^2 = 2$, so $\mathbf{u}.\mathbf{u} = 0$ if $q = 2^e$.

The equation $\mathbf{v}H^{\mathrm{T}} = \mathbf{0}$ defining C can be interpreted as stating that C consists of all the vectors orthogonal to the rows of H, or equivalently, orthogonal to all the vectors in D. Thus C is the *orthogonal code* $D^\perp = \{\mathbf{v} \in V \mid \mathbf{v}.\mathbf{w} = 0 \text{ for all } \mathbf{w} \in D\}$ of D, and interchanging the roles of C and D we see that

$$D = C^\perp = \{\mathbf{w} \in V \mid \mathbf{v}.\mathbf{w} = 0 \text{ for all } \mathbf{v} \in C\}.$$

Thus linear codes come in dual pairs: a generator matrix for one is a parity-check matrix for the other. Note, however, that some linear codes are self-dual, that is $C = C^\perp$: the binary repetition code $\mathcal{R}_2$ is a simple example. More generally, we have:

Example 7.14

Let $q = 2$, let $n = 2m$, and let C be the linear code with basis vectors $\mathbf{u}_i = \mathbf{e}_{2i-1} + \mathbf{e}_{2i}$ for $i = 1, \ldots, m$. Since $\mathbf{u}_i.\mathbf{u}_j = 0$ for all i and j, we have $C \subseteq C^\perp$. Comparing dimensions, we see that $C = C^\perp$.

Example 7.15

The repetition code $\mathcal{R}_n$ is spanned by $\mathbf{1} = 1 \ldots 1$, so

$$\mathcal{R}_n^\perp = \{\mathbf{w} \in \mathcal{V} \mid \mathbf{1}.\mathbf{w} = 0\} = \{\mathbf{w} \in \mathcal{V} \mid w_1 + \cdots + w_n = 0\} = \mathcal{P}_n,$$

and similarly

$$\begin{aligned}
\mathcal{P}_n^\perp &= \{\mathbf{w} \in \mathcal{V} \mid (\mathbf{e}_i - \mathbf{e}_n).\mathbf{w} = 0 \text{ for } i = 1, \ldots, n-1\} \\
&= \{\mathbf{w} \in \mathcal{V} \mid w_i = w_n \text{ for } i = 1, \ldots, n-1\} \\
&= \mathcal{R}_n.
\end{aligned}$$

We have seen that

$$\begin{pmatrix}
1 & & & & -1 \\
& 1 & & & -1 \\
& & \ddots & & \vdots \\
& & & 1 & -1
\end{pmatrix}$$

is a generator matrix for $\mathcal{P}_n$ and also a parity-check matrix for $\mathcal{R}_n$, while

$$\begin{pmatrix} 1 & 1 & \cdots & 1 \end{pmatrix}$$

is a generator matrix for $\mathcal{R}_n$ and a parity-check matrix for $\mathcal{P}_n$.

Example 7.16

The code $\mathcal{H}_7^\perp$ is a linear $[7,3]$-code over F_2; a generator matrix for this code is the parity-check matrix

$$\begin{pmatrix}
0 & 0 & 0 & 1 & 1 & 1 & 1 \\
0 & 1 & 1 & 0 & 0 & 1 & 1 \\
1 & 0 & 1 & 0 & 1 & 0 & 1
\end{pmatrix}$$

for $\mathcal{H}_7$. By taking linear combinations of the rows, we see that the seven non-zero elements of this code all have weight 4, so $d = 4$.

We conclude with a general criterion for determining which matrices are parity-check matrices for a given code. In the statement, 0 denotes a matrix with all entries zero.

Lemma 7.17

Let $\mathcal{C}$ be a linear $[n,k]$-code over F with generator matrix G, and let H be a matrix over F with n columns and $n-k$ rows. Then H is a parity-check matrix for $\mathcal{C}$ if and only if H has rank $n-k$ and satisfies $GH^{\mathrm{T}} = 0$.

Proof

The rows of H form $n - k$ vectors in $\mathcal{V}$, and $GH^{\mathrm{T}} = 0$ if and only if these rows are orthogonal to those of G, or equivalently lie in $\mathcal{C}^{\perp}$. Now H has rank $n - k$ if and only if its rows are linearly independent, or equivalently form a basis for $\mathcal{C}^{\perp}$; thus H satisfies the given conditions if and only if it is a generator matrix for $\mathcal{C}^{\perp}$, that is, a parity-check matrix for $\mathcal{C}$. $\qquad \square$

7.2 Equivalence of Linear Codes

A vector space does not, in general, have a unique basis, so the generator matrix G and the parity-check matrix H of a linear code $\mathcal{C}$ are not generally unique. It is useful to choose them to have as simple a form as possible, for instance to have many zero entries so that calculations are made easier.

The rows $\mathbf{r}_1, \ldots, \mathbf{r}_k$ of G, regarded as elements of $\mathcal{V}$, form a basis for $\mathcal{C}$. The elementary row operations consist of permuting rows, multiplying a row by a non-zero constant, and replacing a row $\mathbf{r}_i$ with $\mathbf{r}_i + a\mathbf{r}_j$ where $j \neq i$ and $a \neq 0$. These may change the basis for $\mathcal{C}$, but they do not change the subspace $\mathcal{C}$ spanned by the rows. We may therefore apply any sequence of these operations to G, giving a new generator matrix for the same code $\mathcal{C}$.

If we permute the *columns* of G, however, we may change $\mathcal{C}$, but the new code will differ from $\mathcal{C}$ only in the order of symbols within code-words; the two codes will have the same parameters such as n, k, d, t, M, R etc., so they are not essentially different. This motivates the following definition. Two linear codes $\mathcal{C}_1$ and $\mathcal{C}_2$ are *equivalent* if they have generator matrices G_1 and G_2 which differ only by elementary row operations and permutations of columns. (Notice that we are *not* allowed to multiply a column by a constant, or to add a multiple of one column to another.) This means that $\mathcal{C}_2$ can be obtained from $\mathcal{C}_1$ by simultaneously re-ordering the symbols in each code-word of $\mathcal{C}_1$. Informally, one tends to think of $\mathcal{C}_1$ and $\mathcal{C}_2$ as "being the same code", even though they generally consist of different code-words.

By systematically using elementary row operations and column permutations, one can convert any generator matrix into the form

$$
G = (\,I_k \mid P\,) = \begin{pmatrix} 1 & & & & * & * & \cdots & * \\ & 1 & & & * & * & \cdots & * \\ & & \ddots & & \vdots & \vdots & & \vdots \\ & & & 1 & * & * & \cdots & * \end{pmatrix}, \tag{7.2}
$$

where I_k is the $k \times k$ identity matrix, and P is a matrix with k rows and $n - k$ columns, represented by the asterisks. We then say that G (or C) is in *systematic form*. In this case, each $\mathbf{a} = a_1 \ldots a_k \in F^k$ is encoded as

$$\mathbf{u} = \mathbf{a}G = a_1 \ldots a_k a_{k+1} \ldots a_n ,$$

where $a_1, \ldots, a_k$ are information digits and $a_{k+1} \ldots a_n = \mathbf{a}P$ is a block of $n - k$ check digits. The information digits are completely arbitrary, whereas the check digits are uniquely determined by $\mathbf{a}$ and G, and are easily calculated as the symbols in $\mathbf{a}P$.

Example 7.18

The generator matrices G for the codes $\mathcal{R}_n$ and $\mathcal{P}_n$ in §7.1 are in systematic form.

Example 7.19.

The generator matrix

$$G_1 = \begin{pmatrix} 1 & 1 & 1 & 0 & 0 & 0 & 0 \\ 1 & 0 & 0 & 1 & 1 & 0 & 0 \\ 0 & 1 & 0 & 1 & 0 & 1 & 0 \\ 1 & 1 & 0 & 1 & 0 & 0 & 1 \end{pmatrix}$$

for $\mathcal{H}_7$, given in §7.1, is not in systematic form. However, by permuting the columns we obtain a generator matrix G_2 for an equivalent code, which is in systematic form: for instance, the permutation $\pi = (1\,7\,4\,5\,2\,6\,3)$, sending column 1 to position 7, etc., gives

$$G_2 = \begin{pmatrix} 1 & 0 & 0 & 0 & 0 & 1 & 1 \\ 0 & 1 & 0 & 0 & 1 & 0 & 1 \\ 0 & 0 & 1 & 0 & 1 & 1 & 0 \\ 0 & 0 & 0 & 1 & 1 & 1 & 1 \end{pmatrix} .$$

Exercise 7.3

The matrices G_1 and G_2 in Example 7.19 generate equivalent codes $\mathcal{C}_1$ and $\mathcal{C}_2$. Are these codes equal or distinct?

If we have a generator matrix

$$G = (I_k \mid P)$$

in systematic form for a linear code $\mathcal{C}$, then we can find a parity-check matrix

$$H = (-P^{\mathrm{T}} \mid I_{n-k}) \tag{7.3}$$

for C. This is the systematic form for a parity-check matrix. We can justify this by using Lemma 7.17: the matrix H shown here has $n - k$ rows and n columns, the presence of the identity matrix I_{n-k} ensures that its rows are independent, so it has rank $n - k$, and finally multiplying block matrices gives

$$GH^{\mathrm{T}} = I_k(-P) + PI_{n-k} = -P + P = 0 .$$

If q is a power of 2 then $a + a = 2a = 0$ for all $a \in F$, so $-a = a$ and we can omit the minus signs, writing H more simply as

$$H = (P^{\mathrm{T}} \mid I_{n-k}) .$$

Example 7.20

In Example 7.1 we found a generator matrix G in systematic form for $\mathcal{R}_n$, where

$$P = (1 \quad 1 \quad \ldots \quad 1)$$

(with $n - 1$ entries), so we get a parity-check matrix

$$H = \begin{pmatrix} -1 & 1 & & \\ -1 & & 1 & \\ \vdots & & & \ddots \\ -1 & & & & 1 \end{pmatrix}$$

for $\mathcal{R}_n$ (with $n - 1$ rows). This is not the same as the parity-check matrix given in Example 7.11, but it results in an equivalent set of parity-check equations for $\mathcal{R}_n$, namely $-v_1 + v_i = 0$ for $i = 2, \ldots, n$.

Example 7.21

The generator matrix given in Example 7.2 for $\mathcal{P}_n$ is in systematic form. It has

$$P = (-1 \quad -1 \quad \cdots \quad -1)^{\mathrm{T}}$$

(with $n - 1$ entries), so $\mathcal{P}_n$ has a parity-check matrix

$$H = (1 \quad 1 \quad \cdots \quad 1 \quad 1)$$

(with n entries).

Example 7.22

The systematic generator matrix given in Example 7.19 for $\mathcal{H}_7$ has

$$P = \begin{pmatrix} 0 & 1 & 1 \\ 1 & 0 & 1 \\ 1 & 1 & 0 \\ 1 & 1 & 1 \end{pmatrix} .$$

Here $q = 2$, so ignoring the minus signs we see that $\mathcal{H}_7$ has a parity-check matrix

$$H = \begin{pmatrix} 0 & 1 & 1 & 1 & 1 & 0 & 0 \\ 1 & 0 & 1 & 1 & 0 & 1 & 0 \\ 1 & 1 & 0 & 1 & 0 & 0 & 1 \end{pmatrix}$$

in systematic form. (Strictly speaking, these are generator and parity-check matrices for a code which is equivalent to $\mathcal{H}_7$, since they have been obtained from those for $\mathcal{H}_7$ by permuting columns; however, as remarked earlier, we will not generally distinguish between equivalent codes.)

Using generator matrices in systematic form, we have an alternative proof of the Singleton bound (Exercise 6.18) for linear codes:

Theorem 7.23

If C is a linear code of length n, dimension k, and minimum distance d, then

$$d \leq 1 + n - k.$$

Proof

By using an equivalent code, we may assume that C has a generator matrix $G = (I_k \mid P)$ in systematic form. Then each row of G is a non-zero code-word of weight at most $1 + n - k$: it has exactly one non-zero information digit (in I_k), and $n - k$ check digits (in P), so at most $1 + n - k$ of its digits are non-zero. It follows from Lemma 6.8 that $d \leq 1 + n - k$. □

Example 7.24

The Singleton bound is attained by $\mathcal{R}_n$, with $k = 1$ and $d = n$, and by $\mathcal{P}_n$, with $k = n - 1$ and $d = 2$; however $\mathcal{H}_7$, with $d = 3$ and $1 + n - k = 4$, does not attain it.

Corollary 7.25

A t-error-correcting linear $[n, k]$-code requires at least $2t$ check digits.

Proof

There are $n - k$ check digits, and Theorems 7.23 and 6.10 give $n - k \geq d - 1 \geq 2t$. □

Example 7.26

The linear codes $\mathcal{R}_3$ and $\mathcal{H}_7$ both have $t = 1$; the number of check digits is $n - k = 2$ or 3 respectively.

7.3 Minimum Distance of Linear Codes

In this section, we will show how the minimum distance of a linear code may be obtained from a parity-check matrix.

Theorem 7.27

Let $\mathcal{C}$ be a linear code of minimum distance d, and let H be a parity-check matrix for $\mathcal{C}$. Then d is the minimum number of linearly dependent columns of H.

Proof

By Lemma 7.10, a vector $\mathbf{v} = v_1 \ldots v_n \in \mathcal{V}$ is a code-word if and only if $\mathbf{v} H^{\mathrm{T}} = \mathbf{0}$, or equivalently $\sum_i v_i \mathbf{c}_i = \mathbf{0}$ where $\mathbf{c}_1, \ldots, \mathbf{c}_n$ are the columns of H. If $\mathbf{v} \neq \mathbf{0}$ then this is a relation of linear dependence between the columns, and conversely any such relation corresponds to a non-zero code-word $\mathbf{v}$; the number of columns appearing in this equation is the number of non-zero terms v_i, which is the weight of $\mathbf{v}$. Thus the least number of linearly dependent columns of H is equal to the least weight of any non-zero code-word, and by Lemma 6.8 this is d. □

Before looking at some examples, let us see what it means for one or two columns of H to be linearly dependent. A single column $\mathbf{c}_i$ is linearly dependent if $v_i \mathbf{c}_i = \mathbf{0}$ for some non-zero $v_i \in F$; multiplying through by v_i^{-1} (which exists since F is a field) we see that this is equivalent to $\mathbf{c}_i = \mathbf{0}$, so Theorem 7.27 tells us that $d = 1$ if and only if H has $\mathbf{0}$ as a column. Two columns $\mathbf{c}_i$ and $\mathbf{c}_j$ (with $i \neq j$) are linearly dependent if and only if $v_i \mathbf{c}_i + v_j \mathbf{c}_j = \mathbf{0}$ with v_i, v_j not both 0; if $\mathbf{c}_i, \mathbf{c}_j \neq \mathbf{0}$ we must have both $v_i, v_j \neq 0$, so we can write this condition as $\mathbf{c}_i = a \mathbf{c}_j$ where $a = -v_j / v_i \in F \setminus \{0\}$. Thus two non-zero columns are linearly dependent if and only if each is a multiple of the other. (In particular, if $q = 2$ then the only possibility for $a \neq 0$ is $a = 1$, so in the binary case two non-zero columns are linearly dependent if and only if they are equal.) It follows from Theorem 7.27 that $d \geq 3$ if and only if the columns of H are non-zero and none is a multiple of any other; in the binary case, this simplifies to the condition that the columns of H are non-zero and distinct from each other.

Example 7.28

The parity-check matrix $H = (1 \quad 1 \quad \cdots \quad 1)$ for $\mathcal{P}_n$ has its columns non-zero and equal, so $\mathcal{P}_n$ has minimum distance $d = 2$.

Example 7.29

In the parity-check matrix

$$H = \begin{pmatrix} 1 & & & & -1 \\ & 1 & & & -1 \\ & & \ddots & & \vdots \\ & & & 1 & -1 \end{pmatrix}$$

for $\mathcal{R}_n$, any set of $n-1$ columns are linearly independent, while $c_1 + \cdots + c_n = 0$ (corresponding to the code-word $\mathbf{1} = 11\ldots1 \in \mathcal{R}_n$), so in this case $d = n$.

Example 7.30

The binary Hamming code $\mathcal{H}_7$ has a parity-check matrix

$$H = \begin{pmatrix} 0 & 0 & 0 & 1 & 1 & 1 & 1 \\ 0 & 1 & 1 & 0 & 0 & 1 & 1 \\ 1 & 0 & 1 & 0 & 1 & 0 & 1 \end{pmatrix}.$$

The columns are non-zero and distinct, so $d \geq 3$; equivalently, there are no code-words of weight 1 or 2. However, $c_1 + c_2 + c_3 = 0$, so there are three linearly dependent columns c_1, c_2 and c_3, corresponding to the fact that $\mathbf{v} = 1110000$ is a code-word of weight 3. (It is the basis element $\mathbf{u}_1$ for $\mathcal{H}_7$ in Example 6.5.) Thus $\mathcal{H}_7$ has minimum distance $d = 3$.

We can use parity-check matrices to give an alternative proof of the Singleton bound (Theorem 7.23) for linear codes. If H is a parity-check matrix for a linear $[n, k]$-code $\mathcal{C}$, then its $n - k$ rows are linearly independent; now the row-rank of any matrix is equal to its column-rank, so H has a set of $n - k$ independent columns, while every set of $n - k + 1$ columns are linearly dependent. By Theorem 7.27 we therefore have $d \leq n - k + 1$.

Corollary 7.31

There is a t-error-correcting linear $[n, k]$-code over F if and only if there is an $(n-k) \times n$ matrix H over F, of rank $n - k$, with every set of $2t$ columns linearly independent.

Proof

($\Rightarrow$) Given such a code $\mathcal{C}$, let H be a parity-check matrix for $\mathcal{C}$, so H has n columns and $n - k$ independent rows. By Theorem 6.10, $\mathcal{C}$ has minimum distance $d \geq 2t + 1$, and by Theorem 7.27, every set of at most $d - 1$ columns are linearly independent, so every set of $2t$ columns are linearly independent.

($\Leftarrow$) Given such a matrix H, let $V = F^n$ and let $C = \{\mathbf{v} \in V \mid \mathbf{v}H^T = \mathbf{0}\}$, a linear code over F of length n. Since H has rank $n - k$, its $n - k$ rows are linearly independent, so C has dimension k. By hypothesis, every set of linearly dependent columns of H contains at least $2t + 1$ columns, so Theorem 7.27 implies that C has minimum distance $d \geq 2t + 1$, and hence C corrects t errors by Theorem 6.10. $\square$

We will give some illustrations of this in the next two sections, where we construct the Hamming and Golay codes.

7.4 The Hamming Codes

The Hamming $[7, 4]$-code $\mathcal{H}_7$ is a 1-error-correcting perfect binary code, of rate $R = \frac{4}{7}$. It is, in fact, one of an infinite sequence of 1-error-correcting perfect binary codes, which have rate R approaching 1 as their length n increases. These codes were introduced by Hamming in 1950 [Ha50], though Golay independently found them around the same time (see [Th83] for a discussion of the priority for this discovery).

For a 1-error-correcting binary linear code, we put $t = 1$ and $q = 2$ in the sphere-packing bound (Corollary 6.17), so the condition for a perfect code becomes

$$2^{n-k} = 1 + \binom{n}{1} = 1 + n.$$

If we put $c = n - k$ (the number of check digits), then this condition is equivalent to

$$n = 2^c - 1. \tag{7.4}$$

Since $k = n - c = 2^c - 1 - c$, the possible values of n and k are as follows:

$c =$	1	2	3	4	5	$\cdots$
$n =$	1	3	7	15	31	$\cdots$
$k =$	0	1	4	11	26	$\cdots$

We now try to construct codes with these parameters. Putting $t = 1$, we see from Corollary 7.31 that such a code C exists if and only if there is a $c \times n$ matrix H over F_2, of rank c, with every pair of columns linearly independent; since $F = F_2 = \{0, 1\}$, this means that the columns $\mathbf{c}_i$ of H must be non-zero and distinct, so H must consist of $n = 2^c - 1$ distinct non-zero column-vectors of length c. Now there are only 2^c distinct binary vectors of length c, so there is no choice about the columns c_i: they must consist of all $2^c - 1$ non-zero binary vectors of length c, in some order (see Example 7.13 for the case $c = 3$). These column-vectors include the c standard basis vectors, which are linearly independent, so such a matrix H has rank c. This proves that C exists, and that any two linear codes with these parameters are equivalent (under a permutation of columns); we call C the *binary Hamming code* $\mathcal{H}_n$ of length $n = 2^c - 1$. (Strictly speaking, we have contructed a *set* of codes of length n here, but since they are all equivalent to each other one tends to refer to them informally as a single code $\mathcal{H}_n$.)

Example 7.32

We will ignore the trivial case $c = 1$, since $\mathcal{H}_1$ consists of the single code-word 0 of length 1. When $c = 2$ we have $n = 3$ and

$$ H = \begin{pmatrix} 0 & 1 & 1 \\ 1 & 0 & 1 \end{pmatrix}; $$

thus $\mathcal{H}_3$ consists of the binary words $\mathbf{v} = v_1 v_2 v_3$ satisfying $v_2 + v_3 = v_1 + v_3 = 0$, or equivalently $v_1 = v_2 = v_3$, so this is the binary repetition code $\mathcal{R}_3$. When $c = 3$ we have the Hamming code $\mathcal{H}_7$ considered earlier. For $c \geq 4$ we obtain infinitely many new perfect codes $\mathcal{H}_n$ of length $n = 2^c - 1$. These codes have rate

$$ R = \frac{k}{n} = \frac{2^c - 1 - c}{2^c - 1} \to 1 $$

as $c \to \infty$, but they correct only one error, so $\mathrm{Pr}_E \nrightarrow 0$ (see Exercise 7.4).

Exercise 7.4

The binary Hamming code $\mathcal{H}_n$ is used, where Γ is a BSC with $P > \frac{1}{2}$, and Δ is nearest neighbour decoding; find Pr_E (see Exercise 6.6 for the case $n = 7$). What happens to Pr_E as $n \to \infty$?

Nearest neighbour decoding with $\mathcal{H}_n$ is very easy. Since $\mathcal{H}_n$ is perfect, with $t = 1$, it corrects every error-pattern $\mathbf{e}$ of weight at most 1. Suppose that $\mathbf{u} \in \mathcal{H}_n$ is transmitted, and $\mathbf{v} = \mathbf{u} + \mathbf{e}$ is received, where $\mathrm{wt}\,(\mathbf{e}) \leq 1$; thus either $\mathbf{e} = \mathbf{0}$, or $\mathbf{e}$ is a standard basis vector $\mathbf{e}_i$ of $\mathcal{V}$. The receiver computes $\mathbf{s} = \mathbf{v}H^{\mathrm{T}}$, called the syndrome[1] of $\mathbf{v}$. Now $\mathbf{v}H^{\mathrm{T}} = (\mathbf{u} + \mathbf{e})H^{\mathrm{T}} = \mathbf{u}H^{\mathrm{T}} + \mathbf{e}H^{\mathrm{T}} = \mathbf{e}H^{\mathrm{T}}$ (since $\mathbf{u}H^{\mathrm{T}} = \mathbf{0}$ by Lemma 7.10), and this is $\mathbf{0}$ or $\mathbf{c}_i^{\mathrm{T}}$ as $\mathbf{e} = \mathbf{0}$ or $\mathbf{e}_i$. If $\mathbf{s} = \mathbf{0}$ the receiver decodes $\mathbf{v}$ as $\Delta(\mathbf{v}) = \mathbf{v}\,(= \mathbf{u})$, and if $\mathbf{s} = \mathbf{c}_i^{\mathrm{T}}$ then $\Delta(\mathbf{v}) = \mathbf{v} - \mathbf{e}_i$, formed by changing the i-th symbol of $\mathbf{v}$. This method always decodes correctly

---

[1] The terminology is from Medicine, where the syndrome is an indication of what is wrong with the patient.

if wt $(\mathbf{e}) \leq 1$, but never if wt $(\mathbf{e}) > 1$: in this case $\mathbf{v} = \mathbf{u}' + \mathbf{e}'$ for some unique $\mathbf{u}' \in \mathcal{H}_n$ with wt $(\mathbf{e}') \leq 1$, where $\mathbf{u}' \neq \mathbf{u}$, and the above algorithm decodes $\mathbf{v}$ incorrectly as $\Delta(\mathbf{v}) = \mathbf{u}'$.

Example 7.33

Let us use $\mathcal{H}_7$, with parity-check matrix

$$H = \begin{pmatrix} 0 & 1 & 1 & 1 & 1 & 0 & 0 \\ 1 & 0 & 1 & 1 & 0 & 1 & 0 \\ 1 & 1 & 0 & 1 & 0 & 0 & 1 \end{pmatrix}$$

in systematic form. Suppose that $\mathbf{u} = 1101001 \in \mathcal{H}_7$ is sent, and $\mathbf{v} = 1101101 \in V$ is received, so the error-pattern is $\mathbf{e} = \mathbf{e}_5$. The syndrome is $\mathbf{s} = \mathbf{v}H^{\mathrm{T}} = 100$, which is the transpose $\mathbf{c}_5^{\mathrm{T}}$ of the fifth column of H; this indicates an error in the fifth position, so changing this entry of $\mathbf{v}$ we get $\Delta(\mathbf{v}) = 1101001 = \mathbf{u}$. On the other hand, suppose that $\mathbf{v}' = 1001101$ is received, with error-pattern $\mathbf{e}' = \mathbf{e}_2 + \mathbf{e}_5$; in this case the syndrome is $\mathbf{s}' = 001 = \mathbf{c}_7^{\mathrm{T}}$, indicating an error in the seventh position, so $\Delta(\mathbf{v}') = 1001100 \neq \mathbf{u}$. Thus, instead of correcting two errors, the code has actually created a third.

Exercise 7.5

Use the parity-check matrix in Example 7.33 to verify that $\mathbf{u} = 1100110$ is a code-word in $\mathcal{H}_7$. Suppose that this word $\mathbf{u}$ is sent, and $\mathbf{v} = 1000110$ is received. Find the syndrome, and hence find $\Delta(\mathbf{v})$. Investigate what happens if $\mathbf{v}' = 0000110$ is received, and give an explanation.

Decoding with $\mathcal{H}_n$ is particularly simple if we order the columns $\mathbf{c}_i$ of H so that the row vectors $\mathbf{c}_1^{\mathrm{T}}, \ldots, \mathbf{c}_n^{\mathrm{T}}$ are the binary representations of the integers $i = 1, \ldots, n$, in that order. Thus for $n = 7$ we take

$$H = \begin{pmatrix} 0 & 0 & 0 & 1 & 1 & 1 & 1 \\ 0 & 1 & 1 & 0 & 0 & 1 & 1 \\ 1 & 0 & 1 & 0 & 1 & 0 & 1 \end{pmatrix},$$

as in Example 7.13; this is equivalent under the permutation $(1\,3\,6\,2\,5\,4\,7)$ of the columns to the parity-check matrix used in Example 7.33. A syndrome $\mathbf{s} = \mathbf{0}$ is interpreted as $\mathbf{e} = \mathbf{0}$, that is, no error, while a non-zero syndrome $\mathbf{s}$ is the binary representation of the position i where a single error $\mathbf{e}_i$ has appeared.

Example 7.34

Let us use the equivalent version of $\mathcal{H}_7$ defined by the above parity-check matrix. Applying the permutation $(1\,3\,6\,2\,5\,4\,7)$ to the word 1101001 used in Example 7.33, we obtain the codeword $\mathbf{u} = 1010101 \in \mathcal{H}_7$. If this is transmitted and $\mathbf{v} = 1010001$ is received, then the syndrome is $\mathbf{s} = \mathbf{v}H^{\mathrm{T}} = 101$; this is the binary representation of the integer 5, so changing the fifth entry of $\mathbf{v}$ we obtain $\Delta(\mathbf{v}) = 1010101 = \mathbf{u}$.

Exercise 7.6

Use the equivalent version of $\mathcal{H}_7$, as in Example 7.34, to investigate what happens when $\mathbf{u} = 0111100$ is sent and $\mathbf{v} = 0011100$ is received.

There is a similar construction of perfect 1-error-correcting linear codes for prime-powers $q > 2$: we take the columns of H to be

$$n = \frac{q^c - 1}{q - 1} = 1 + q + q^2 + \cdots + q^{c-1} \tag{7.5}$$

pairwise linearly independent vectors of length c over F_q (this is the maximum number possible: see Exercise 7.7). The resulting linear code has length n, dimension $k = n - c$, and minimum distance $d = 3$, so $t = 1$ (see Exercise 7.7 again). As in the binary case, $R \to 1$ as $c \to \infty$, but $\mathrm{Pr_E} \not\to 0$.

Exercise 7.7

Show that if $\mathcal{W} = F_q^c$ then the maximum number of vectors in $\mathcal{W}$, such that no two of them are linearly dependent, is $(q^c - 1)/(q - 1)$. Show that if any such set of vectors form the columns of a parity-check matrix H, then the resulting linear code over F_q is perfect and 1-error correcting.

Example 7.35

If $q = 3$ and $c = 2$, then $n = 4$ and $k = 2$. We can take

$$H = \begin{pmatrix} 1 & 1 & 1 & 0 \\ 1 & 2 & 0 & 1 \end{pmatrix},$$

giving a perfect 1-error-correcting linear $[4, 2]$-code over F_3.

7.5 The Golay Codes

Golay used Corollary 7.31 to construct the two perfect codes $\mathcal{G}_{11}$ and $\mathcal{G}_{23}$ which now carry his name. In a remarkable paper [Go49], occupying only half a page, he described not just these two codes, but also the perfect binary repetition codes $\mathcal{R}_n$ (n odd), and the perfect codes constructed at the end of §7.4 for all primes q (the extension to prime-powers q came a little later).

Recall from §6.4 that a perfect linear code is one which attains equality in the sphere-packing bound, so

$$\sum_{i=0}^{t} \binom{n}{i}(q-1)^i = q^{n-k}. \tag{7.6}$$

Now Exercise 6.19 suggests that there may be a perfect linear code with $q = 3$, $n = 11$, $k = 6$ and $t = 2$. To construct such a code, Golay considered a parity-check matrix

$$H = \begin{pmatrix} 1 & 1 & 1 & 2 & 2 & 0 & 1 & 0 & 0 & 0 & 0 \\ 1 & 1 & 2 & 1 & 0 & 2 & 0 & 1 & 0 & 0 & 0 \\ 1 & 2 & 1 & 0 & 1 & 2 & 0 & 0 & 1 & 0 & 0 \\ 1 & 2 & 0 & 1 & 2 & 1 & 0 & 0 & 0 & 1 & 0 \\ 1 & 0 & 2 & 2 & 1 & 1 & 0 & 0 & 0 & 0 & 1 \end{pmatrix}$$

over F_3, in systematic form, with $n = 11$ columns and $n - k = 5$ independent rows. With considerable patience, one can show that there are no sets of four or fewer linearly dependent columns, whereas there is a set of five linearly dependent columns (for instance $\mathbf{c}_2 - \mathbf{c}_7 - \mathbf{c}_8 + \mathbf{c}_9 + \mathbf{c}_{10} = \mathbf{0}$). It follows from Theorem 7.27 that the code $\mathcal{C}$ defined by H has $d = 5$, and hence $t = 2$ by Theorem 6.10. Since

$$\sum_{i=0}^{t} \binom{n}{i}(q-1)^i = 1 + \binom{11}{1} \cdot 2 + \binom{11}{2} \cdot 2^2 = 243 = 3^5 = q^{n-k},$$

this code $\mathcal{C}$ is perfect. It is the *ternary Golay code* $\mathcal{G}_{11}$ of length 11.

Similarly, taking $q = 2$, $n = 23$ and $k = 12$ (as suggested by Exercise 6.19), Golay used a binary parity-check matrix $H = (P^T \mid I_{11})$ where

$$P^T = \begin{pmatrix} 1 & 0 & 0 & 1 & 1 & 1 & 0 & 0 & 0 & 1 & 1 & 1 \\ 1 & 0 & 1 & 0 & 1 & 1 & 0 & 1 & 1 & 0 & 0 & 1 \\ 1 & 0 & 1 & 1 & 0 & 1 & 1 & 0 & 1 & 0 & 1 & 0 \\ 1 & 0 & 1 & 1 & 1 & 0 & 1 & 1 & 0 & 1 & 0 & 0 \\ 1 & 1 & 0 & 0 & 1 & 1 & 1 & 0 & 1 & 1 & 0 & 0 \\ 1 & 1 & 0 & 1 & 0 & 1 & 1 & 1 & 0 & 0 & 0 & 1 \\ 1 & 1 & 0 & 1 & 1 & 0 & 0 & 1 & 1 & 0 & 1 & 0 \\ 1 & 1 & 1 & 0 & 0 & 1 & 0 & 1 & 0 & 1 & 1 & 0 \\ 1 & 1 & 1 & 0 & 1 & 0 & 1 & 0 & 0 & 0 & 1 & 1 \\ 1 & 1 & 1 & 1 & 0 & 0 & 0 & 0 & 1 & 1 & 0 & 1 \\ 0 & 1 & 1 & 1 & 1 & 1 & 1 & 1 & 1 & 1 & 1 & 1 \end{pmatrix}.$$

An even more tedious process shows that the minimum number of linearly dependent columns of H is seven, so the corresponding code, the *binary Golay code* $\mathcal{G}_{23}$ of length 23, has $d = 7$ and hence $t = 3$. Again, this code is perfect, since

$$\sum_{i=0}^{t} \binom{n}{i}(q-1)^i = 1 + \binom{23}{1} + \binom{23}{2} + \binom{23}{3} = 2048 = 2^{11} = q^{n-k}.$$

The *extended Golay codes* $\mathcal{G}_{12} = \overline{\mathcal{G}}_{11}$ and $\mathcal{G}_{24} = \overline{\mathcal{G}}_{23}$ are linear $[12, 6]$- and $[24, 12]$-codes over F_3 and F_2. Although they are not perfect, they are very important examples of codes, having links with many mathematical structures such as Steiner systems, lattices, sphere-packings and simple groups [CL91, CS92]. For a fascinating account of the origin of the Golay codes, and of their connections with some of these topics, see [Th83]. Because of these links, there are many alternative ways of constructing the Golay codes; most of them are more enlightening than Golay's original construction, outlined above, though none are completely straightforward (see Exercises 7.17 and 7.18 for two relatively simple examples, based on results in [CL91]). Here we will show how the binary Golay codes may be obtained from combinatorial objects called Steiner systems.

If S is any set with n elements, then the power set

$$\mathcal{P}(S) = \{U \mid U \subseteq S\}$$

is an n-dimensional vector space over F_2, in which the sum $U + V$ of two subsets U and V is their symmetric difference $(U \cup V) \setminus (U \cap V)$, and the zero element is the empty set $\emptyset$. If $S = \{s_1, \ldots, s_n\}$, then each subset U can be represented as a vector $\mathbf{u} = u_1 \ldots u_n \in \mathcal{V} = F_2^n$, with $u_i = 1$ or 0 as $s_i \in U$ or $s_i \notin U$. We can therefore regard any non-empty subset $\mathcal{C} \subseteq \mathcal{P}(S)$, that is, any non-empty set of subsets of S, as a binary code of length n, which is linear if and only if it is closed under addition. We have $\text{wt}(\mathbf{u}) = |U|$ and $d(\mathbf{u}, \mathbf{v}) = |U + V|$, so to achieve a large minimum distance d we choose $\mathcal{C}$ so that distinct subsets $U, V \in \mathcal{C}$ are sufficiently different from each other.

One systematic way to do this is to use block designs. A *t-design* on a set S is a set of subsets of S, called *blocks*, all of the same size, such that each set of t elements of S are contained in the same number λ of blocks. These regularity conditions impose strong restrictions on the resulting codes. The connections between designs and codes are explained in detail in [CL91], and here we will restrict attention to some simple examples.

Writing l in place of the traditional t (which we have already used for the number of errors corrected), we define a *Steiner system* to be an *l-design* with $\lambda = 1$, that is, a set of m-element blocks B in an n-element set S, such that each set of l elements of S are contained in a unique block. We will denote such a system by $S(l, m, n)$.

Example 7.36

Let S be the set of all 1-dimensional subspaces of the vector space $W = F^c$, where $c \geq 2$, so $|S| = (q^c - 1)/(q - 1)$. Each 2-dimensional subspace of W contains $q + 1$ elements of S, and we regard these $(q+1)$-element subsets as the blocks. Each pair of distinct 1-dimensional subspaces of W generate a unique 2-dimensional subspace, so each pair of elements of S lie in a unique block. We therefore have a Steiner system with $l = 2$, $m = q + 1$ and $n = (q^c - 1)/(q - 1)$; this is the *projective geometry* $PG(c - 1, q)$, with the lines of this geometry as blocks. In Fig. 7.1 we see the seven 3-element blocks of the *Fano plane* $PG(2, 2)$; see Exercise 7.12 for the connection between these geometries and the Hamming codes.

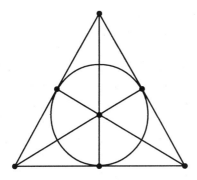

Figure 7.1

If a Steiner system $S(l, m, n)$ has b blocks, then

$$\binom{n}{l} = b \binom{m}{l}.$$

This is because each of the $\binom{n}{l}$ l-element subsets of S lies in a unique block, and each of the b blocks contains $\binom{m}{l}$ such subsets. Thus $\binom{m}{l}$ divides $\binom{n}{l}$, imposing a restriction on the possible parameters l, m and n. In fact there are further restrictions. If $s \in S$, then it is easy to check that $S' = S \setminus \{s\}$ is a Steiner system $S(l - 1, m - 1, n - 1)$, in which the blocks are the sets $B \setminus \{s\}$ where B is a block of S containing s; the preceding argument then implies that $\binom{m-1}{l-1}$ divides $\binom{n-1}{l-1}$. By iterating this we can obtain further restrictions.

Example 7.37

If S is a Steiner system $S(2, 3, n)$, then these two conditions state that 3 divides $n(n - 1)/2$ and 2 divides $n - 1$, so $n \equiv 1$ or 3 mod (6). In fact, this necessary condition for the existence of S is also known to be sufficient [Ha67, Theorem 15.4.3], with $PG(c - 1, 2)$ providing examples for $n = 2^c - 1$.

Example 7.38

The triple $(5, 8, 24)$ satisfies the above necessary conditions, so there could conceivably be a Steiner system $S(5, 8, 24)$, with $b = \binom{24}{5}/\binom{8}{5} = 759$ blocks. Such a system S has been shown to exist, and to be essentially unique; its automorphism group (the set of permutations of S taking blocks to blocks) is the *Mathieu group* M_{24}, a simple group of order $244, 823, 040$. Now let $\mathcal{C}$ be the subspace of $\mathcal{V} = \mathcal{P}(S) = F_2^{24}$ spanned by the blocks of S. Assuming only the definition of a Steiner system, without needing to know the blocks, one can use simple counting arguments to show that $\mathcal{C}$ consists of:

    1 set of size 0, namely $\emptyset$;

    759 sets B of size 8, namely the blocks;

    2576 sets $B + B'$ of size 12, where B and B' are blocks with $|B \cap B'| = 2$;

    759 sets $B + B'$ of size 16, where B and B' are disjoint blocks;

    1 set of size 24, namely S, the sum of three disjoint blocks.

(See §7.3 of [An74] for the details.) Now $1 + 759 + 2576 + 759 + 1 = 4096 = 2^{12}$, so $\mathcal{C}$ is a binary linear $[24, 12]$-code. This is the extended Golay code $\mathcal{G}_{24}$. The code-words have weights 0, 8, 12, 16 and 24, so $\mathcal{G}_{24}$ has minimum distance $d = 8$. By puncturing $\mathcal{G}_{24}$ at any single position (deleting the i-th symbol from all code-words, for some fixed i), we obtain a binary linear $[23, 12]$-code with $d = 7$, and this is the perfect Golay code $\mathcal{G}_{23}$; the choice of i here is unimportant, since all the resulting codes are equivalent.

Exercise 7.8

Prove that in a Steiner system $S = S(5, 8, 24)$, every element $s \in S$ lies in 253 blocks, every two elements lie in 77 blocks, every three elements lie in 21 blocks, and every four elements lie in 5 blocks.

One can reverse the argument, and obtain the Steiner system from the code: the blocks of $S(5, 8, 24)$ are the supports $U = \{i \mid u_i \neq 0\}$ of the code-words $\mathbf{u} \in \mathcal{G}_{24}$ of weight 8 (see [CL91] for this approach). Similarly the supports of the code-words of weight 7 in $\mathcal{G}_{23}$ form a Steiner system $S(4, 7, 23)$, while the words of weight 5 in $\mathcal{G}_{11}$ and 6 in $\mathcal{G}_{12}$ yield Steiner systems $S(4, 5, 11)$ and $S(5, 6, 12)$. In these last two cases, however, as in most non-binary cases, the derivation of the code from the design is more complicated.

7.6 The Standard Array

For nearest neighbour decoding, given any received word $\mathbf{v} \in \mathcal{V}$ we need to be able to find the code-word $\mathbf{u} = \Delta(\mathbf{v}) \in \mathcal{C}$ nearest to $\mathbf{v}$. When $\mathcal{C}$ is a linear code there is an algorithm for doing this based on the *standard array*, which is essentially a table in which the elements of $\mathcal{V}$ are arranged into cosets of the subspace $\mathcal{C}$.

Suppose that
$$\mathcal{C} = \{\mathbf{u}_1, \mathbf{u}_2, \ldots, \mathbf{u}_M\}$$
is a linear code with $M = q^k$ elements; $\mathbf{0}$ must be a code-word, so we will choose the numbering so that $\mathbf{u}_1 = \mathbf{0}$. For $i = 1, 2, 3, \ldots$ we form the i-th row of the standard array by first choosing $\mathbf{v}_i$ to be an element of $\mathcal{V}$, not in any previous row, of least possible weight (so, in particular, $\mathbf{v}_1 = \mathbf{0}$); we then let the i-th row consist of the elements of the coset

$$\mathbf{v}_i + \mathcal{C} = \{\mathbf{v}_i + \mathbf{u}_1 \ (= \mathbf{v}_i), \ \mathbf{v}_i + \mathbf{u}_2, \ \ldots, \ \mathbf{v}_i + \mathbf{u}_M\}$$

of $\mathcal{C}$ in $\mathcal{V}$, written in that order. Thus the first row is $\mathbf{v}_1 + \mathcal{C} = \mathbf{0} + \mathcal{C} = \mathcal{C}$, distinct rows are disjoint, and the process stops after $q^n/M = q^{n-k}$ rows have been formed, one for each coset. When this happens, every $\mathbf{v} \in \mathcal{V}$ appears exactly once in the array as

$$\mathbf{v} = \mathbf{v}_i + \mathbf{u}_j, \tag{7.7}$$

for some i and j, so that $\mathbf{v}$ is the j-th term in the i-th row. The elements $\mathbf{v}_i$ are coset representatives for $\mathcal{C}$ in $\mathcal{V}$, called *coset leaders*. By construction, we have

$$\mathrm{wt}(\mathbf{v}_1) \leq \mathrm{wt}(\mathbf{v}_2) \leq \mathrm{wt}(\mathbf{v}_3) \leq \ldots \ ;$$

we draw a horizontal line across the array, immediately under the last row to satisfy $\mathrm{wt}(\mathbf{v}_i) \leq t$, where $t = \lfloor \frac{d-1}{2} \rfloor$ is the number of errors corrected by $\mathcal{C}$. Note that the standard array is not generally unique: there are usually several possible vectors $\mathbf{v}_i$ which can be chosen as coset leader for the i-th row.

Example 7.39

Let $\mathcal{C}$ be the binary repetition code $\mathcal{R}_4$ of length $n = 4$, so $q = 2$, $k = 1$ and the code-words are $\mathbf{u}_1 = \mathbf{0} = 0000$ and $\mathbf{u}_2 = \mathbf{1} = 1111$. There are $q^{n-k} = 8$ cosets of $\mathcal{C}$ in $\mathcal{V}$, each consisting of two vectors, so the standard array has eight rows and two columns. We are forced to choose $\mathbf{v}_1 = \mathbf{0}$ as the first coset leader; the next four are the standard basis vectors (the only words of weight 1) in some order, and the last three (which are not uniquely determined) have weight 2. This code has $d = 4$, so $t = 1$ and hence we draw the line under the fifth row. For instance, a possible form for the standard array is:

$$
\begin{array}{ll}
0000 & 1111 \\
1000 & 0111 \\
0100 & 1011 \\
0010 & 1101 \\
0001 & 1110 \\
\hline
1100 & 0011 \\
1010 & 0101 \\
1001 & 0110
\end{array}
$$

Lemma 7.40

(a) If $\mathbf{v}$ is in the j-th column of the standard array (that is, $\mathbf{v} = \mathbf{v}_i + \mathbf{u}_j$ for some i), then $\mathbf{u}_j$ is a nearest code-word to $\mathbf{v}$.

(b) If, in addition, $\mathbf{v}$ is above the line in the standard array (that is, $\mathrm{wt}(\mathbf{v}_i) \leq t$), then $\mathbf{u}_j$ is the unique nearest code-word to $\mathbf{v}$.

Proof

(a) Let $\mathbf{v} = \mathbf{v}_i + \mathbf{u}_j$, and suppose that $\mathbf{u}_j$ is not a nearest code-word to $\mathbf{v}$, so $d(\mathbf{v}, \mathbf{u}_{j'}) < d(\mathbf{v}, \mathbf{u}_j)$ for some $\mathbf{u}_{j'} \in C$. Since $d(\mathbf{v}, \mathbf{u}) = \mathrm{wt}(\mathbf{v} - \mathbf{u})$ and $\mathbf{v} - \mathbf{u}_j = \mathbf{v}_i$ we have

$$
\mathrm{wt}(\mathbf{v} - \mathbf{u}_{j'}) < \mathrm{wt}(\mathbf{v} - \mathbf{u}_j) = \mathrm{wt}(\mathbf{v}_i);
$$

now

$$
\mathbf{v} - \mathbf{u}_{j'} = \mathbf{v}_i + \mathbf{u}_j - \mathbf{u}_{j'} \in \mathbf{v}_i + C
$$

(since $\mathbf{u}_j - \mathbf{u}_{j'} \in C$), which contradicts the choice of $\mathbf{v}_i$ as an element of least weight in its coset in the construction of the standard array.

(b) In addition to the above, let $\mathrm{wt}(\mathbf{v}_i) \leq t$, and suppose that $d(\mathbf{v}, \mathbf{u}_{j'}) \leq d(\mathbf{v}, \mathbf{u}_j)$ for some $\mathbf{u}_{j'} \in C$. Then

$$
\begin{aligned}
d(\mathbf{u}_j, \mathbf{u}_{j'}) &\geq d & \text{(by definition of } d\text{)} \\
&> 2t & \text{(by Theorem 6.10)} \\
&\geq 2d(\mathbf{v}, \mathbf{u}_j) & \text{(since } \mathrm{wt}(\mathbf{v}_i) \leq t\text{)} \\
&\geq d(\mathbf{v}, \mathbf{u}_j) + d(\mathbf{v}, \mathbf{u}_{j'}) & \text{(since } d(\mathbf{v}, \mathbf{u}_j) \geq d(\mathbf{v}, \mathbf{u}_{j'})\text{)} \\
&\geq d(\mathbf{u}_j, \mathbf{u}_{j'}) & \text{(by the triangle inequality),}
\end{aligned}
$$

so $d(\mathbf{u}_j, \mathbf{u}_{j'}) > d(\mathbf{u}_j, \mathbf{u}_{j'})$, which is impossible. $\qquad\square$

This shows that the sphere $S_t(\mathbf{u}_j)$ of radius t about $\mathbf{u}_j$, defined in §6.4, is the part of the j-th column above the line. Thus $\mathcal{C}$ is perfect if and only if the entire standard array is above the line.

We can use Lemma 7.40 for decoding. Suppose that a code-word $\mathbf{u} \in \mathcal{C}$ is transmitted, and $\mathbf{v} = \mathbf{u} + \mathbf{e} \in \mathcal{V}$ is received, where $\mathbf{e}$ is the error-pattern. The receiver finds $\mathbf{v} = \mathbf{v}_i + \mathbf{u}_j$ in the standard array, and decides that $\Delta(\mathbf{v}) = \mathbf{u}_j$ was most likely to have been sent, since this is a nearest neighbour of $\mathbf{v}$ in $\mathcal{C}$ (indeed, it is *the* nearest neighbour if $\mathbf{v}$ is above the line). Thus each received word $\mathbf{v}$ is decoded as the code-word $\mathbf{u}_j$ heading its column in the standard array. This decision is correct if and only if $\mathbf{u} = \mathbf{u}_j$, that is, if and only if $\mathbf{e} = \mathbf{v}_i$, so this rule gives correct decoding if and only if the error-pattern is a coset leader.

Example 7.41

Let $\mathcal{C} = \mathcal{R}_4$, with the standard array as in Example 7.39. Suppose that $\mathbf{u} = 1111$ is sent, and the error-pattern is $\mathbf{e} = 0100$ (so only the second symbol of $\mathbf{u}$ is transmitted incorrectly). Then $\mathbf{v} = 1111 + 0100 = 1011$ is received, and since this is in the column of the array headed by $\mathbf{u}_2 = 1111$, the receiver decides (correctly) that $\Delta(\mathbf{v}) = 1111$ was sent. However, if the error-pattern is $\mathbf{e} = 0110$ then $\mathbf{v} = 1111 + 0110 = 1001$ is received; this is in the column headed by $\mathbf{u}_1 = 0000$, so the receiver decides (incorrectly) that $\Delta(1001) = 0000$ was sent. In fact, this choice of array corrects all error-patterns of weight 0 or 1, but only the patterns $\mathbf{e} = 1100, 1010$ and 1001 of weight 2, and none of weight 3 or 4. Any choice of array will correct three error-patterns of weight 2, but not necessarily these three.

The advantage of this method of decoding is that it is relatively simple to understand and to implement. The disadvantages are that it requires a great deal of storage (the standard array contains every word in $\mathcal{V}$), and searching for received words $\mathbf{v}$ in the array could be time-consuming. In the next section, we therefore consider an equivalent but more efficient method of decoding linear codes.

7.7 Syndrome Decoding

Syndrome decoding is a more streamlined version of the decoding algorithm described in §7.6. If H is a parity-check matrix for a linear code $\mathcal{C} \subseteq \mathcal{V}$, then the *syndrome* of a vector $\mathbf{v} \in \mathcal{V}$ is the vector

$$\mathbf{s} = \mathbf{v}H^{\mathrm{T}} \in F^{n-k} \tag{7.8}$$

(we used this idea in §7.4 in connection with the binary Hamming codes). Thus $\mathbf{s} = s_1 \ldots s_{n-k}$, where $s_i = \mathbf{v}.\mathbf{r}_i$ and $\mathbf{r}_i$ is the i-th row of H, so s_i is the result of applying the i-th parity-check condition to $\mathbf{v}$. The next result shows that the syndrome $\mathbf{s}$ allows us to decide which coset of $\mathcal{C}$ contains $\mathbf{v}$, or equivalently which row of the standard array contains $\mathbf{v}$. Recall first that two vectors $\mathbf{v}, \mathbf{v}' \in V$ lie in the same coset of the subspace $\mathcal{C}$ (that is, $\mathbf{v} + \mathcal{C} = \mathbf{v}' + \mathcal{C}$) if and only if $\mathbf{v} - \mathbf{v}' \in \mathcal{C}$.

Lemma 7.42

Let $\mathcal{C}$ be a linear code, with parity-check matrix H, and let $\mathbf{v}, \mathbf{v}' \in V$ have syndromes $\mathbf{s}, \mathbf{s}'$. Then $\mathbf{v}$ and $\mathbf{v}'$ lie in the same coset of $\mathcal{C}$ if and only if $\mathbf{s} = \mathbf{s}'$.

Proof

We have

$$
\begin{aligned}
\mathbf{v} + \mathcal{C} = \mathbf{v}' + \mathcal{C} &\iff \mathbf{v} - \mathbf{v}' \in \mathcal{C} \\
&\iff (\mathbf{v} - \mathbf{v}')H^{\mathrm{T}} = \mathbf{0} \qquad \text{(by Lemma 7.10)} \\
&\iff \mathbf{v}H^{\mathrm{T}} = \mathbf{v}'H^{\mathrm{T}} \\
&\iff \mathbf{s} = \mathbf{s}'.
\end{aligned}
$$

$\square$

This shows that a vector $\mathbf{v} \in V$ lies in the i-th row of the standard array if and only if it has the same syndrome as $\mathbf{v}_i$, that is, $\mathbf{v}H^{\mathrm{T}} = \mathbf{v}_iH^{\mathrm{T}}$. We therefore form a *syndrome table*, consisting of two columns: the coset leaders $\mathbf{v}_i$ (chosen as in §7.6) are in the first column, and their syndromes $\mathbf{s}_i = \mathbf{v}_iH^{\mathrm{T}}$ are opposite them in the second column.

Example 7.43

Let $\mathcal{C}$ be the binary repetition code $\mathcal{R}_4$, with standard array as given in Example 7.39, so the coset leaders $\mathbf{v}_i$ are the words in its first column. If we use the parity-check matrix

$$
H = \begin{pmatrix} 1 & & -1 \\ & 1 & -1 \\ & & 1 & -1 \end{pmatrix} = \begin{pmatrix} 1 & & 1 \\ & 1 & 1 \\ & & 1 & 1 \end{pmatrix}
$$

given in Example 7.11, and apply it to any vector $\mathbf{v} = v_1 v_2 v_3 v_4 \in V$, then we find that $\mathbf{s} = \mathbf{v}H^{\mathrm{T}} = s_1 s_2 s_3$ where $s_i = v_i + v_4$ for $i = 1, 2, 3$. Applying this to the coset leaders $\mathbf{v} = \mathbf{v}_1, \ldots, \mathbf{v}_8$, we obtain the corresponding syndromes $\mathbf{s}_i$. This gives the following syndrome table:

$\mathbf{v}_i$	$\mathbf{s}_i$
0000	000
1000	100
0100	010
0010	001
0001	111
1100	110
1010	101
1001	011

In general, if we have a parity-check matrix H and a syndrome table for a linear code $\mathcal{C}$, then decoding proceeds as follows. Given any received word $\mathbf{v}$, we compute its syndrome $\mathbf{s} = \mathbf{v}H^{\mathrm{T}}$, and then find $\mathbf{s}$ in the second column of the syndrome table, say $\mathbf{s} = \mathbf{s}_i$, the i-th entry. If $\mathbf{v}_i$ is the coset leader corresponding to $\mathbf{s}_i$ in the table, then Lemma 7.42 implies that $\mathbf{v}$ lies in the same coset of $\mathcal{C}$ as $\mathbf{v}_i$, so $\mathbf{v} = \mathbf{v}_i + \mathbf{u}_j$ for some code-word $\mathbf{u}_j$. As in §7.6, we therefore decode $\mathbf{v}$ as $\mathbf{u}_j$. Thus

$$\Delta(\mathbf{v}) = \mathbf{u}_j = \mathbf{v} - \mathbf{v}_i, \quad \text{where} \quad \mathbf{v}H^{\mathrm{T}} = \mathbf{s}_i.$$

Example 7.44

Let $\mathcal{C} = \mathcal{R}_4$ again, with parity-check matrix H and syndrome table as in Example 7.43. If $\mathbf{v} = 1101$ is received, we first compute its syndrome $\mathbf{s} = \mathbf{v}H^{\mathrm{T}} = 001$. This is $\mathbf{s}_4$ in the syndrome table, so we decode $\mathbf{v}$ as

$$\Delta(\mathbf{v}) = \mathbf{v} - \mathbf{v}_4 = 1101 - 0010 = 1111.$$

The advantage of this method is that, once H is known and the syndrome table has been constructed, decoding is relatively quick: given $\mathbf{v}$, the syndrome $\mathbf{s} = \mathbf{v}H^{\mathrm{T}}$ is easily computed; since the syndrome table is much smaller than the standard array, $\mathbf{s}$ can generally be found in it much faster than $\mathbf{v}$ can be found in the standard array, especially if the syndromes are arranged in some convenient order; finally, subtracting $\mathbf{v}_i$ from $\mathbf{v}$ to give $\mathbf{u}_j$ is easy.

Example 7.45

We can reinterpret the decoding algorithm for Hamming codes, described in §7.4, in terms of syndrome tables. If $\mathcal{C}$ is a binary Hamming code $\mathcal{H}_n$, then the coset leaders $\mathbf{v}_i$ are the $n + 1$ vectors $\mathbf{v} \in V$ of weight $\mathrm{wt}(\mathbf{v}) \leq 1$, starting with $\mathbf{v}_1 = \mathbf{0}$ and followed by the n standard basis vectors of V is some order. Their syndromes $\mathbf{s}_i$ consist of $\mathbf{s}_1 = \mathbf{0}$, followed by the transposes of the columns of H in the corresponding order. Let us take these columns to be the binary representations of the integers $1, 2, \ldots, n$, in that order (as in §7.4), and let us order the non-zero coset leaders by taking $\mathbf{v}_{i+1} = \mathbf{e}_i$ for $i = 1, \ldots, n$; then the syndromes $\mathbf{s}_1, \ldots, \mathbf{s}_{n+1}$ are the binary representations of the integers $0, 1, \ldots, n$, in that order. If a received vector $\mathbf{v}$ produces a syndrome $\mathbf{s} = \mathbf{0}$, this is interpreted as meaning that no error has occurred, so $\Delta(\mathbf{v}) = \mathbf{v}$; on the other hand a syndrome $\mathbf{s} \neq \mathbf{0}$ is interpreted as the binary representation of the position i where a single error has occurred, so $\Delta(\mathbf{v}) = \mathbf{v} - \mathbf{e}_i$.

Exercise 7.9

Let $\mathcal{C}$ be the binary linear code spanned by $011011, 101101$ and 111000. Find a generator matrix G for $\mathcal{C}$ in systematic form, and hence find a parity-check matrix H for $\mathcal{C}$. Find the code-word $\mathbf{c}$ with information digits 110, and verify that $\mathbf{c}H^{\mathrm{T}} = \mathbf{0}$. Find the rate R and the minimum distance d of $\mathcal{C}$. Find a syndrome table for $\mathcal{C}$; which error patterns does it correct? Find Pr_{E}, where the channel Γ is a BSC with $P > \frac{1}{2}$.

7.8 Supplementary Exercises

Exercise 7.10

Show that the number of distinct k-dimensional linear codes $\mathcal{C} \subseteq \mathcal{V} = F_q^n$ is

$$\frac{(q^n - 1)(q^n - q) \ldots (q^n - q^{n-k+1})}{(q^k - 1)(q^k - q) \ldots (q^k - q^{k-1})}.$$

Exercise 7.11

Show that if L_1 and L_2 are distinct lines in the Fano plane $S = PG(2, 2)$, then their symmetric difference $L_1 + L_2$ is the complement of a third line. Deduce that the subspace $\mathcal{C}$ of $\mathcal{V} = \mathcal{P}(S) \cong F_2^7$ spanned by the lines consists of $\emptyset$, the 7 lines, their 7 complements, and S. Show that this code is equivalent to the Hamming code $\mathcal{H}_7$.

Exercise 7.12

Show that if $\mathcal{C}$ is any perfect 1-error-correcting binary code of length n, then the supports in $S = \{1, \ldots, n\}$ of the code-words of weight 3 are the blocks of a Steiner system $S(2,3,n)$ on S. Show that if $C = \mathcal{H}_n$, where $n = 2^c - 1$, the resulting Steiner system is isomorphic to $PG(c-1,2)$.

Exercise 7.13

An *automorphism* of a code $C \subseteq V = F^n$ is a permutation of the n coordinates which maps the set C to itself. Show that these form a subgroup $\text{Aut}(C)$ of the symmetric group S_n. What are $\text{Aut}(\mathcal{R}_n)$ and $\text{Aut}(\mathcal{P}_n)$? List the code-words of the binary code $\mathcal{R}_2 \oplus \mathcal{R}_2$ (see Exercise 6.20) and hence find $|\text{Aut}(\mathcal{R}_2 \oplus \mathcal{R}_2)|$. Show that the number of distinct codes in V equivalent to C is $n!/|\text{Aut}(C)|$, and find all such codes when $C = \mathcal{R}_2 \oplus \mathcal{R}_2$.

Exercise 7.14

Show that $\mathcal{H}_7$ and $PG(2,2)$ both have a group of automorphisms isomorphic to the group $GL(3,2)$ of 3×3 invertible matrices over F_2. How many automorphisms does $\mathcal{H}_7$ have, and how many distinct codes $C \subseteq F_2^7$ are equivalent to $\mathcal{H}_7$? What are the corresponding results for $\mathcal{H}_n$, where $n = 2^c - 1$?

Exercise 7.15

Show that if C is a perfect t-error-correcting binary code of length n, then the supports in $S = \{1, \ldots, n\}$ of the code-words of weight $d = 2t+1$ are the blocks of a Steiner system $S(t+1, d, n)$ on S. Deduce that $\binom{d-i}{t+1-i}$ divides $\binom{n-i}{t+1-i}$ for $i = 0, 1, \ldots, t$.

Exercise 7.16

What does the factorisation of $1 + 90 + \binom{90}{2}$ suggest about the possible existence of a perfect binary code of length $n = 90$? Prove that such a perfect code cannot exist.

Exercise 7.17

Show that if $\mathbf{u}, \mathbf{v}$ are binary vectors, then $\text{wt}(\mathbf{u} + \mathbf{v}) = \text{wt}(\mathbf{u}) + \text{wt}(\mathbf{v}) - 2c(\mathbf{u}, \mathbf{v})$, where $c(\mathbf{u}, \mathbf{v})$ is the number of i such that $u_i = v_i = 1$. Let G be the block matrix $(I_{12} \mid P)$, where the 12 rows and columns of P are indexed by the vertices of an icosahedron, with $P_{ij} = 0$ if ij is an edge, and $P_{ij} = 1$ otherwise. Show that G is a generator matrix for a self-dual binary linear $[24,12]$-code C, and that $G' = (P \mid I_{12})$ is also a generator matrix for C. Show that every code-word has weight divisible by 4, but none has weight 4, and hence C has minimum distance 8. Show that the punctured code C° is a 3-error-correcting perfect binary $[23,12]$-code. (It can be shown that C and C° are equivalent to the Golay codes $\mathcal{G}_{24}$ and $\mathcal{G}_{23}$.)

Exercise 7.18

Let H be the parity-check matrix given in Example 7.35 for the ternary Hamming $[4, 2]$-code, and let

$$G = \begin{pmatrix} J + I & I & I \\ O & H & -H \end{pmatrix}$$

where I and J are the 4×4 identity and all-ones matrices. Show that G is a generator matrix for a ternary $[12, 6]$-code C of minimum distance 6, and that the punctured code C° is a perfect 2-error-correcting ternary $[11, 6]$-code. (It can be shown that C and C° are equivalent to the Golay codes $\mathcal{G}_{12}$ and $\mathcal{G}_{11}$.)

Exercise 7.19

The r-th order *Reed–Muller code* $\mathcal{RM}(r, m)$ of length $n = 2^m$ can be defined inductively as follows: $\mathcal{RM}(0, m)$ is the binary repetition code of length n, $\mathcal{RM}(m, m) = F_2^n$, and if $0 < r < m$ then

$$\mathcal{RM}(r, m) = \mathcal{RM}(r, m - 1) * \mathcal{RM}(r - 1, m - 1)$$

(where $*$ is defined in Exercise 6.20). Show that $\mathcal{RM}(r, m)$ is a binary linear code of length $n = 2^m$, dimension $k = \sum_{i=0}^{r} \binom{m}{i}$ and minimum distance $d = 2^{m-r}$. List all the code-words in $\mathcal{RM}(1, 2)$ and $\mathcal{RM}(1, 3)$. Find a generator matrix G and hence a parity-check matrix H for $\mathcal{RM}(1, 3)$; using H, verify that this code has minimum distance 4.

Exit, pursued by a bear. (*The Winter's Tale*)

Suggestions for Further Reading

Shannon's classic 1948 paper [Sh48] has been published, with a non-technical introduction by Weaver, as a short book [SW63], and is well worth reading. Ash [As65] gives a precise and detailed mathematical account of Information Theory, while Reza's approach [Re61] is principally aimed at engineers, as is McEliece's rather sophisticated treatment of Information and Coding Theory in [McE77]. Chambers [Ch85] and Jones [Jo79] give concise introductions to Information Theory from a more applied point of view than we have taken, while Welsh [We88] emphasises the connections with Cryptography.

In Coding Theory, Hill [Hi86] and Pless [Pl82] both continue the development of the subject somewhat further than we have, but starting at a similar level. There are rather more advanced texts by Berlekamp [Be68], Blake and Mullin [BM75, BM76], Pretzel [Pr92] and van Lint [Li82], while the standard reference books on Coding Theory are the encyclopædic works by MacWilliams and Sloane [MS77] and by Pless and Huffman [PH98]. Thompson [Th83] provides a very readable account of the early history of coding theory, in particular the Hamming and Golay codes and their connections with sphere-packings and simple groups; Anderson [An74] gives a good introduction to the combinatorial background for these links, including the Steiner system $S(5, 8, 24)$, while Conway and Sloane [CS92] give a much deeper and more detailed treatment of this material. Connections between codes, graphs and block designs are explored in detail by Cameron and van Lint [CL91]. Applications of algebraic geometry to codes are discussed by Pretzel [Pr92], Goppa [Go88], and van Lint and van der Geer [LG88]; Stichtenoth [St93] gives a sophisticated account of the closely-related subject of algebraic function fields and their connections with Coding Theory.

Variable-length codes, as studied in Chapter 1, can be regarded as purely algebraic objects. The set T^* of all words in some alphabet T is a monoid, which means that it has a binary operation (concatenation) which satisfies the associative law $u(vw) = (uv)w$ and has an identity element (the empty word ε). Unique decodability of a code $C \subseteq T^*$ is equivalent to the condition that C should be a set of free generators for the submonoid of T^* which it generates. These and other similar links between codes and algebra are explored in great detail by Berstel and Perrin in [BP85].

Trees, which we introduced in Chapter 1 to describe certain classes of codes, such as instantaneous codes, are important both in graph theory and (especially in the case of binary trees) in other areas such as computer science. Huffman's algorithm is one of a number of tree algorithms discussed in some detail by Knuth in [Kn73]. For other applications of Huffman's algorithm, see [De74], [Ev79], [Kn73], [ST81], [Zi59].

Entropy, introduced in Chapter 3, also plays an important role in thermodynamics as a measure of the disorganisation of a system, with p_i the probability that the system is in the i-th state of its phase space. The Second Law of Thermodynamics states that entropy cannot decrease, thus providing a direction for time by showing that systems tend towards disorder. Brillouin discusses the connections between information theory and thermodynamics in [Br56]. There are also strong links between entropy and ergodic theory (the theory of measure-preserving transformations): see Billingsley [Bi65], for instance.

The basic probability theory required in this book is covered in most textbooks on that subject. The more advanced Law of Large Numbers is used in Chapter 5 to prove Shannon's Fundamental Theorem, and is explained in Appendix B; for further discussion, and a proof, we recommend Feller [Fe50]. Similarly, there are numerous textbooks covering the linear algebra we need in Chapters 6 and 7, Blyth and Robertson [BR98] being a good example. We also use a few results from analysis and calculus, such as the Mean Value Theorem and Stirling's approximation for $n!$; Fisher [Fi83] and Lang [La83] are typical of a number of good undergraduate references for these topics. Finite fields are used in Chapters 6 and 7; for further background and applications, one could consult [KR83].

There is a lot to be said for reading original papers, in order to get a feeling for how the originators of a subject thought and expressed themselves. This is particularly true in the area of Information and Coding Theory: most of these papers, being relatively recent, are easily available, and many can be read without a deep mathematical background. The collections of key papers edited by Berlekamp [Be74] and Slepian [Sl74] cover most of the important developments in the first 25 years of the subject. Readers with library access to periodicals such as *Bell System Technical Journal*, *IEEE Transactions in Information Theory*, and *Information and Control* will also find a number of interesting and important research papers: Shannon's paper [Sh48], for instance, a very influential paper on variable-length codes by Gilbert and Moore [GM59], results by Schwartz [Sc64] and Golomb [Go80] on the non-uniqueness of Huffman codes, and the paper by Kelley [Ke56] on gambling from which Exercises 5.11 and 5.12 are derived.

Appendix *A*
Proof of the Sardinas–Patterson Theorem

The Sardinas–Patterson Theorem was stated without a complete proof as Theorem 1.10 in §1.2. Recall that a code $C \subseteq T^*$ is defined to be uniquely decodable if, whenever $u_1 \ldots u_l = v_1 \ldots v_m$ with $u_i, v_j \in C$, then $l = m$ and $u_i = v_i$ for each i. Given a code C, we define $C_0 = C$,

$$C_n = \{\, w \in T^+ \mid uw = v \text{ where } u \in C,\, v \in C_{n-1} \text{ or } u \in C_{n-1},\, v \in C \,\}$$

for each $n \geq 1$, and $C_\infty = \bigcup_{n=1}^{\infty} C_n$. Then the Theorem states that C is uniquely decodable if and only if $C \cap C_\infty = \emptyset$. The proof we give here is based on those given by Bandyopadhyay [Ba63] and Seeley [Se67].

First we need some notation. If $u, v \in T^*$ and u is a prefix of v, that is, $v = uw$ for some $w \in T^*$, we write $u \leq v$; we also write $w = u^{-1}v$, meaning that w is obtained by deleting the prefix u from v. (Note that u^{-1} alone is not defined.) If, in addition, $u \neq v$ then we write $u < v$. Now we can start the proof.

($\Leftarrow$) If C is not uniquely decodable, we can choose an ambiguous code-sequence of least length, say

$$u_1 \ldots u_l = v_1 \ldots v_m, \qquad\qquad (*)$$

where $u_i, v_j \in C$, and if $l = m$ then $u_i \neq v_i$ for some i. We will work from left to right through the word represented by $(*)$, using the overlapping code-words u_i and v_j to define a sequence of words $x_n \in C_n$ for $n = 1, 2, \ldots$, until eventually some x_n coincides with either u_l or v_m, giving us the required element of $C \cap C_\infty$.

By minimality we have $u_1 \neq v_1$ (otherwise $u_2 \ldots u_l = v_2 \ldots v_m$ is a shorter ambiguous code-sequence), so $(*)$ implies that either $u_1 < v_1$ or $v_1 < u_1$;

renaming if necessary, we may assume that $v_1 < u_1$. Then the non-empty word $x_1 = v_1^{-1} u_1$ is in $\mathcal{C}_1$, since $v_1 x_1 = u_1$ with $u_1, v_1 \in \mathcal{C} = \mathcal{C}_0$. If $v_1 v_2 < u_1$, then since $v_2 \in \mathcal{C}$ and $v_2(v_1 v_2)^{-1} u_1 = v_1^{-1} u_1 = x_1 \in \mathcal{C}_1$ the word $x_2 = (v_1 v_2)^{-1} u_1$ is in $\mathcal{C}_2$. We continue like this until we reach the largest integer i_1 such that $v_1 \ldots v_{i_1} < u_1$; note that $x_n = (v_1 \ldots v_n)^{-1} u_1$ is in $\mathcal{C}_n$ for $1 \le n \le i_1$. This is illustrated in Fig. A.1, where horizontal segments denote words.

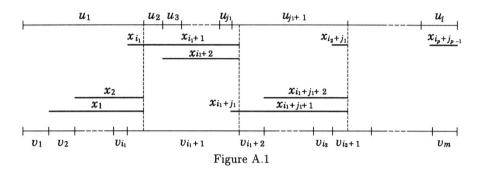

Figure A.1

At the next stage, we must have $u_1 \le v_1 \ldots v_{i_1+1}$. If $u_1 = v_1 \ldots v_{i_1+1}$ (so that $l = 1$ and $m = i_1 + 1$ by minimality), then $v_{i_1+1} = x_{i_1} \in \mathcal{C} \cap \mathcal{C}_{i_1} \subseteq \mathcal{C} \cap \mathcal{C}_\infty$, so $\mathcal{C} \cap \mathcal{C}_\infty \ne \emptyset$. Hence we may assume that $u_1 < v_1 \ldots v_{i_1+1}$. Then the word $x_{i_1+1} = u_1^{-1} v_1 \ldots v_{i_1+1}$ is in $\mathcal{C}_{i_1+1}$, since $x_{i_1} x_{i_1+1} = v_{i_1+1} \in \mathcal{C}$ and $x_{i_1} \in \mathcal{C}_{i_1}$. If $u_1 u_2 < v_1 \ldots v_{i_1+1}$, then the word $x_{i_1+2} = (u_1 u_2)^{-1}(v_1 \ldots v_{i_1+1})$ is in $\mathcal{C}_{i_1+2}$ since $u_2 x_{i_1+2} = x_{i_1+1} \in \mathcal{C}_{i_1+1}$ and $u_2 \in \mathcal{C}$. Again we continue like this until we reach the largest integer j_1 such that $u_1 \ldots u_{j_1} < v_1 \ldots v_{i_1+1}$, giving $x_{i_1+j_1} = (u_1 \ldots u_{j_1})^{-1}(v_1 \ldots v_{i_1+1}) \in \mathcal{C}_{i_1+j_1}$.

Now $v_1 \ldots v_{i_1+1} \le u_1 \ldots u_{j_1+1}$, and if we have equality here then (as before) we have reached the end of our minimal ambiguous code-sequence (∗), with $u_{j_1+1} = x_{i_1+j_1} \in \mathcal{C} \cap \mathcal{C}_\infty$. Assuming that $v_1 \ldots v_{i_1+1} < u_1 \ldots u_{j_1+1}$, we have $x_{i_1+j_1+1} = (v_1 \ldots v_{i_1+1})^{-1}(u_1 \ldots u_{j_1+1})$ in $\mathcal{C}_{i_1+j_1+1}$ since $x_{i_1+j_1} x_{i_1+j_1+1} = u_{j_1+1} \in \mathcal{C}$ with $x_{i_1+j_1} \in \mathcal{C}_{i_1+j_1}$. We continue like this until we reach the largest integer i_2 such that $v_1 \ldots v_{i_2} < u_1 \ldots u_{j_1+1}$. This gives $x_{i_2+j_1} = (v_1 \ldots v_{i_2})^{-1}(u_1 \ldots u_{j_1+1}) \in \mathcal{C}_{i_2+j_1}$.

At the next stage, $u_1 \ldots u_{j_1+1} \le v_1 \ldots v_{i_2+1}$, and we continue as we did when we reached $u_1 \le v_1 \ldots v_{i_1+1}$. Eventually, we must reach the end of the code-sequence (∗). Now $|u_l| \ne |v_m|$, for otherwise $u_l = v_m$ and hence $u_1 \ldots u_{l-1} = v_1 \ldots v_{m-1}$, contradicting minimality. If $|u_l| > |v_m|$ (as in Fig. A.1) then we finish with $m - 1 = i_p$ for some p, and $v_m = x_{i_p+j_{p-1}} \in \mathcal{C} \cap \mathcal{C}_\infty$. If, on the other hand, $|u_l| < |v_m|$ then we end with $l - 1 = j_p$ for some p, and $u_l = x_{i_p+j_p} \in \mathcal{C} \cap \mathcal{C}_\infty$.

Example A.1

Suppose that the minimal ambiguous sequence $(*)$ has the form $u_1 u_2 u_3 = v_1 v_2 v_3 v_4 v_5$, where the words u_i and v_j overlap as in Fig. A.2.

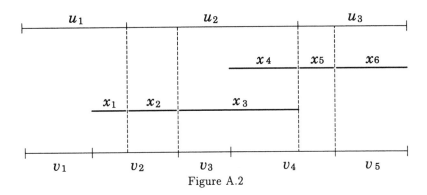

Figure A.2

By following the above process we find that $i_1 = 1$, $j_1 = 1$, $i_2 = 3$, $j_2 = 2$, $i_3 = 4 = m - 1$, so $p = 3$ and $v_5 = x_6 \in C \cap C_6 \subseteq C \cap C_\infty$.

$(\Rightarrow)$ Suppose that $C \cap C_\infty \neq \emptyset$, so $C \cap C_n \neq \emptyset$ for some $n \geq 1$; let v_n denote an element of $C \cap C_n$. Applying the definition of the sets $C_n, \ldots, C_2$ (in that order), we see that the following statement is true for $2 \leq k \leq n$:

(S_k) either $u_{k-1} v_k = v_{k-1}$ or $v_{k-1} v_k = u_{k-1}$, for some $u_{k-1} \in C$, $v_{k-1} \in C_{k-1}$.

Similarly, the definition of C_1 then gives

(S_1) $uv_1 = u'$ for some $u, u' \in C$.

Here each $v_k \neq \varepsilon$, so in particular $u \neq u'$, a fact we need later. These statements (S_k) will enable us to construct a word which can be factorised into code-words in two different ways. To do this, we need to show that for each $k = 1, \ldots, n-1$, the element $v_k \in C_k$ also satisfies the following statement:

(T_k) $v_k y_k = z_k$ for some $y_k, z_k \in C^*$.

Firstly, (T_{n-1}) is true, since (S_n) gives

$$v_{n-1} y_{n-1} = z_{n-1}$$

with either $y_{n-1} = \varepsilon$ and $z_{n-1} = u_{n-1} v_n$, or $y_{n-1} = v_n$ and $z_{n-1} = u_{n-1}$; in either case $y_{n-1}, z_{n-1} \in C^*$ since $u_{n-1}, v_n \in C$. Now we will show that (T_k) implies (T_{k-1}) for $2 \leq k \leq n - 1$. Suppose that (T_k) is true. Statement (S_k) gives

$$u_{k-1} v_k = v_{k-1} \quad \text{or} \quad v_{k-1} v_k = u_{k-1},$$

so either

$$v_{k-1}y_k = u_{k-1}v_ky_k = u_{k-1}z_k \quad \text{or} \quad v_{k-1}z_k = v_{k-1}v_ky_k = u_{k-1}y_k.$$

Both of these assertions have the form

$$v_{k-1}y_{k-1} = z_{k-1},$$

where

$$y_{k-1} = y_k, \ z_{k-1} = u_{k-1}z_k \quad \text{or} \quad y_{k-1} = z_k, \ z_{k-1} = u_{k-1}y_k$$

respectively; in either case, y_{k-1} and z_{k-1} are elements of C^* (since y_k and z_k are, with $u_{k-1} \in C$), so (T_{k-1}) is proved. It follows that each (T_k) is true, so taking $k = 1$ gives

$$v_1y_1 = z_1 \quad \text{for some} \quad y_1, z_1 \in C^*.$$

Then (S_1) implies that

$$u'y_1 = uv_1y_1 = uz_1,$$

where $y_1, z_1 \in C^*$ and u, u' are distinct code-words. Thus the equation $u'y_1 = uz_1$ gives two distinct ways of factorising the same code-sequence into code-words, so C is not uniquely decodable. $\qquad\square$

Example A.2

For an illustration of this, we return to Example 1.12 of §1.2, where $C = \{01, 1, 2, 210\}$. There we found $1 \in C \cap C_3$, so in the above notation we put $n = 3$ and $v_3 = 1$. Then the statements (S_k) take the form

(S_3) $0.1 = 01$, that is, $v_2v_3 = u_2$ where $u_2 = 01 \in C$ and $v_2 = 0 \in C_2$;

(S_2) $1.0 = 10$, that is, $u_1v_2 = v_1$ where $u_1 = 1 \in C$ and $v_1 = 10 \in C_1$;

(S_1) $2.10 = 210$, that is, $uv_1 = u'$ where $u = 2 \in C$ and $u' = 210 \in C$.

    Thus (T_2), that is, $v_2y_2 = z_2$, becomes $0.1 = 01$ where $y_2 = 1 \in C^*$ and $z_2 = 01 \in C^*$. Similarly (T_1), that is, $v_1y_1 = z_1$, becomes $10.1 = 101$ where $y_1 = 1 \in C^*$ and $z_1 = 101 = 1.01 \in C^*$. Using $(S_1), (T_1)$ and the factorisation of z_1 we have

$$210.1 = 2.10.1 = 2.101 = 2.1.01.$$

This gives two factorisations 210.1 and $2.1.01$ of the code-sequence 2101 as a product of code-words, confirming that C is not uniquely decodable.

Appendix *B*
The Law of Large Numbers

In the proof of Shannon's Fundamental Theorem (Theorem 5.9), one needs to estimate the number of errors in a transmitted code-word $\mathbf{u} = u_1 \ldots u_n$ of length n, that is, the number of non-zero coordinates $e_i = v_i - u_i$ of the error-pattern $\mathbf{e} = \mathbf{v} - \mathbf{u}$, where $\mathbf{v} = v_1 \ldots v_n$ is the received word. In the case of the BSC, where $A = B = \mathbf{Z}_2$, we have $e_i = 0$ or 1 as u_i is transmitted correctly or incorrectly. These two events have probabilities P and Q ($= \overline{P}$), independently of what happens to the other digits of $\mathbf{u}$, so one can regard $e_1, \ldots, e_n$ as the outcomes of n successive Bernoulli trials (independent, identically distributed random variables). If we regard the values 0 and 1 of each e_i as real numbers, rather than as integers mod (2), then the number of errors is $\sum_i e_i$. The Law of Large Numbers tells us about the sum (or equivalently the average) of the values of a large number of Bernoulli trials, so it gives us the required estimate for the number of errors.

Let X be a random variable, taking finitely many real values x_j with probabilities p_j, so that $0 \le p_j \le 1$ and $\sum_j p_j = 1$. The mean, or expected value of X is

$$\mu = \mathrm{E}\,(X) = \sum_j p_j x_j.$$

Now let $X_1, \ldots, X_n$ be n successive Bernoulli trials of X, that is, n independent random variables taking the values x_j with the same probabilities p_j as X. (Typical examples are repeatedly tossing the same coin, or rolling the same die.) If

$$Y = \frac{1}{n} \sum_{i=1}^{n} X_i$$

is the average of n outcomes, then our intuition suggests that when n is large, Y should be close to μ. For instance, if X is an unbiased coin, and we score $x_j = 1$ or 0 for heads or tails, then $\mu = \frac{1}{2}$ and we expect that $Y \approx \frac{1}{2}$ also.

Of course, we cannot guarantee that $Y \approx \mu$ in all cases. If we toss the coin $n = 10$ times, then an outcome of 10 heads ($Y = 1$) is unlikely, but not impossible: it has probability $2^{-10} = 1/1024 \approx 0.001$, which is small but non-zero. Even an outcome of, say, 6 heads out of 10 (giving $Y = 0.6$) is not particularly surprising, since it has probability $\binom{10}{6}/2^{10} \approx 0.205$, compared with the probability of about 0.246 for the most likely outcome, 5 heads. If we toss the coin $n = 100$ times, however, then it is far more likely that Y will be close to $\frac{1}{2}$: for instance, $Y = 1$ now has probability $2^{-100} \approx 10^{-30}$, and $Y = 0.6$ has probability $\binom{100}{60}/2^{100} \approx 0.010$, so both are extremely unlikely (though still not impossible!).

The Law of Large Numbers confirms this intuition, telling us that as n increases, it is increasingly likely that Y will be close to μ. More precisely, it states that for any $\eta > 0$, we have $|Y - \mu| \leq \eta$ with probability approaching 1 as $n \to \infty$, or equivalently,

$$\lim_{n \to \infty} \Pr \left(\left| \frac{1}{n} \sum_{i=1}^{n} X_i - \mu \right| > \eta \right) = 0.$$

This is, in fact, more correctly known as the *Weak Law of Large Numbers*, since there are stronger versions of this result. For further details of this and other limit theorems in Statistics, with proofs, see [Fe50].

Appendix C
Proof of Shannon's Fundamental Theorem

In §5.4 we stated Shannon's Fundamental Theorem for the BSC:

Theorem 5.9

Let Γ be a binary symmetric channel with $P > \frac{1}{2}$, so Γ has capacity $C = 1 - H(P) > 0$, and let $\delta, \varepsilon > 0$. Then for all sufficiently large n there is a code $\mathcal{C} \subseteq \mathbf{Z}_2^n$, of rate R satisfying $C - \varepsilon \leq R < C$, such that nearest neighbour decoding gives error-probability $\mathrm{Pr_E} < \delta$.

We will now give a complete proof, filling in the gaps in the outline proof in §5.4.

Proof

Let $\mathcal{V} = \mathbf{Z}_2^n$. We will regard a code $\mathcal{C} \subseteq \mathcal{V}$ as an ordered sequence $(\mathbf{u}_1, \ldots, \mathbf{u}_M)$ of distinct elements of $\mathcal{V}$, so different orderings of the same elements are treated as different codes. This is just a technical device to help the proof along: having shown that an ordered code satisfies the theorem, we can then forget the ordering.

First we consider decoding. Let us choose a small $\eta > 0$ (we will specify later how small), and put $\rho = n(Q + \eta)$ where $Q = \overline{P}$. The motivation for this is that we expect about nQ incorrect symbols in any word of length n, so the transmitted and received words $\mathbf{u}$ and $\mathbf{v}$ probably satisfy $d(\mathbf{u}, \mathbf{v}) \approx nQ$; by taking ρ to be slightly larger than nQ, we can expect that $d(\mathbf{u}, \mathbf{v}) \leq \rho$ with high probability.

159

We will use ρ to find an upper bound for the average error-probability Pr_{E}. Suppose that a code-word $\mathbf{u}_i \in C$ is transmitted, and $\mathbf{v} = \mathbf{u}_i + \mathbf{e} \in V$ is received, where $\mathbf{e}$ is the random error-pattern. If $d(\mathbf{u}_i, \mathbf{v}) \le \rho$, and $d(\mathbf{u}_j, \mathbf{v}) > \rho$ for all $j \ne i$, then nearest neighbour decoding gives $\Delta(\mathbf{v}) = \mathbf{u}_i$, which is correct. Equivalently, if decoding is incorrect then either $d(\mathbf{u}_i, \mathbf{v}) > \rho$ or $d(\mathbf{u}_j, \mathbf{v}) \le \rho$ for some $j \ne i$. Averaging over all $\mathbf{e}$, we deduce that the conditional probability $\mathrm{Pr}\,(\Delta(\mathbf{v}) \ne \mathbf{u}_i \mid \mathbf{u}_i)$ of incorrect decoding, given that $\mathbf{u}_i$ is transmitted, satisfies

$$\mathrm{Pr}\,(\Delta(\mathbf{v}) \ne \mathbf{u}_i \mid \mathbf{u}_i) \le \mathrm{Pr}\,(d(\mathbf{u}_i, \mathbf{u}_i + \mathbf{e}) > \rho) + \sum_{j \ne i} \mathrm{Pr}\,(d(\mathbf{u}_j, \mathbf{u}_i + \mathbf{e}) \le \rho). \quad (C.1)$$

Next we show that the first term on the right-hand side can be made arbitrarily small. Writing $\mathbf{e} = (e_1, \ldots, e_n)$ with each $e_i = 0$ or 1, we have

$$d(\mathbf{u}_i, \mathbf{u}_i + \mathbf{e}) = \mathrm{wt}(\mathbf{e}) = \sum_{k=1}^{n} e_k$$

(where the addition is in $\mathbf{Z}$, not $\mathbf{Z}_2$). Now $\rho = n(Q + \eta)$, so

$$\mathrm{Pr}\,(d(\mathbf{u}_i, \mathbf{u}_i + \mathbf{e}) > \rho) = \mathrm{Pr}\,\left(\frac{1}{n}\sum_{k=1}^{n} e_k > Q + \eta\right)$$

$$\le \mathrm{Pr}\,\left(\left|\frac{1}{n}\sum_{k=1}^{n} e_k - Q\right| > \eta\right).$$

We can regard $e_1, \ldots, e_n$ as Bernoulli trials, taking the values 0 or 1 with probabilities P and Q. The mean, or expected value $\mu = \mathrm{E}(e_k)$ of each e_k is therefore $P.0 + Q.1 = Q$, so the Weak Law of Large Numbers (Appendix B) gives

$$\mathrm{Pr}\,\left(\left|\frac{1}{n}\sum_{k=1}^{n} e_k - Q\right| > \eta\right) \to 0 \quad \text{as} \quad n \to \infty.$$

(This simply says that for large n, the average of $e_1, \ldots, e_n$ is probably close to their mean.) Thus

$$\mathrm{Pr}\,\left(\left|\frac{1}{n}d(\mathbf{u}_i, \mathbf{u}_i + \mathbf{e}) - Q\right| > \eta\right) \to 0 \quad \text{as} \quad n \to \infty,$$

so

$$\mathrm{Pr}\,(d(\mathbf{u}_i, \mathbf{u}_i + \mathbf{e}) > \rho) < \frac{\delta}{2} \quad\quad\quad (C.2)$$

for all sufficiently large n. This deals with the first term in $(C.1)$. Averaging $(C.1)$ over all code-words $\mathbf{u}_i = \mathbf{u}_1, \ldots, \mathbf{u}_M$ of C (assumed to be equiprobable), we see that $\mathcal{C}$ has error-probability

$$
\begin{aligned}
\text{Pr}_E &= \frac{1}{M} \sum_{i=1}^{M} \text{Pr}\left(\Delta(\mathbf{v}) \neq \mathbf{u}_i \mid \mathbf{u}_i\right) \\
&\leq \frac{1}{M} \sum_{i=1}^{M} \left(\text{Pr}\left(d(\mathbf{u}_i, \mathbf{u}_i + \mathbf{e}) > \rho\right) + \sum_{j \neq i} \text{Pr}\left(d(\mathbf{u}_j, \mathbf{u}_i + \mathbf{e}) \leq \rho\right) \right) \\
&< \frac{1}{M} \sum_{i=1}^{M} \left(\frac{\delta}{2} + \sum_{j \neq i} \text{Pr}\left(d(\mathbf{u}_j, \mathbf{u}_i + \mathbf{e}) \leq \rho\right) \right) \\
&= \frac{\delta}{2} + \frac{1}{M} \sum_{i=1}^{M} \sum_{j \neq i} \text{Pr}\left(d(\mathbf{u}_j, \mathbf{u}_i + \mathbf{e}) \leq \rho\right). \quad (C.3)
\end{aligned}
$$

The double sum in $(C.3)$ is difficult to deal with, since it depends heavily on the particular code $\mathcal{C} = (\mathbf{u}_1, \ldots, \mathbf{u}_M)$ chosen: the probabilities tend to be large or small as the code-words are close or far apart. Shannon's brilliant idea was to "even out" this effect by taking an average over all possible choices of $\mathcal{C}$. If $f(\mathcal{C})$ is a number associated with each code $\mathcal{C}$, then the *average* of f is

$$
\overline{f} = \frac{(2^n - M)!}{2^n!} \sum_{\mathcal{C}} f(\mathcal{C}),
$$

where the summation is over all the $2^n!/(2^n - M)!$ distinct M-element ordered codes $\mathcal{C} = (\mathbf{u}_1, \ldots, \mathbf{u}_M)$ in $\mathcal{V}$. (There should be no confusion with the notation $\overline{f} = 1 - f$, which we will not use here.) Applying this to $(C.3)$, we see that the average of the error-probabilities of our codes satisfies

$$
\begin{aligned}
\overline{\text{Pr}_E} &< \frac{\delta}{2} + \frac{1}{M} \overline{\sum_i \sum_{j \neq i} \text{Pr}\left(d(\mathbf{u}_j, \mathbf{u}_i + \mathbf{e}) \leq \rho\right)} \\
&= \frac{\delta}{2} + \frac{1}{M} \sum_i \sum_{j \neq i} \overline{\text{Pr}\left(d(\mathbf{u}_j, \mathbf{u}_i + \mathbf{e}) \leq \rho\right)}; \quad (C.4)
\end{aligned}
$$

in the first line we use the fact that δ and M are independent of $\mathcal{C}$, and in the second the fact that the average of a sum is the sum of their averages.

To deal with the summation in $(C.4)$, let us choose any pair of subscripts $i \neq j$. Now $d(\mathbf{u}_j, \mathbf{u}_i + \mathbf{e}) = d(\mathbf{u}_j - \mathbf{u}_i, \mathbf{e})$, so $d(\mathbf{u}_j, \mathbf{u}_i + \mathbf{e}) \leq \rho$ if and only if $\mathbf{u}_j - \mathbf{u}_i$ lies in $S_\rho(\mathbf{e})$, the sphere of radius ρ centred at $\mathbf{e}$. Let

$$
f_\mathbf{e}(\mathcal{C}) = \begin{cases} 1 & \text{if } \mathbf{u}_j - \mathbf{u}_i \in S_\rho(\mathbf{e}), \\ 0 & \text{if } \mathbf{u}_j - \mathbf{u}_i \notin S_\rho(\mathbf{e}). \end{cases}
$$

Then

$$\overline{\Pr\left(d(\mathbf{u}_j, \mathbf{u}_i + \mathbf{e}) \le \rho\right)} = \overline{\Pr\left(\mathbf{u}_j - \mathbf{u}_i \in S_\rho(\mathbf{e})\right)}$$

$$= \frac{(2^n - M)!}{2^n!} \sum_{\mathcal{C}} \Pr\left(\mathbf{u}_j - \mathbf{u}_i \in S_\rho(\mathbf{e})\right)$$

$$= \frac{(2^n - M)!}{2^n!} \sum_{\mathcal{C}} \left(\sum_{\mathbf{e}} \Pr\left(\mathbf{e}\right) f_{\mathbf{e}}(\mathcal{C}) \right)$$

$$= \sum_{\mathbf{e}} \left(\Pr\left(\mathbf{e}\right) \frac{(2^n - M)!}{2^n!} \sum_{\mathcal{C}} f_{\mathbf{e}}(\mathcal{C}) \right)$$

$$= \sum_{\mathbf{e}} \Pr\left(\mathbf{e}\right) \overline{f_{\mathbf{e}}(\mathcal{C})}, \qquad (C.5)$$

where $\Pr\left(\mathbf{e}\right)$ is the probability $(= P^{n-w}Q^w)$ of a particular error-pattern $\mathbf{e}$ of weight w. (This argument simply says that the operations of averaging over $\mathcal{C}$ and $\mathbf{e}$ commute with each other.) Now for each fixed $\mathbf{e}$, $\overline{f_{\mathbf{e}}(\mathcal{C})}$ is the proportion of codes $\mathcal{C}$ such that $\mathbf{u}_j - \mathbf{u}_i \in S_\rho(\mathbf{e})$. As $\mathcal{C} = (\mathbf{u}_1, \ldots, \mathbf{u}_M)$ ranges over all codes, $\mathbf{u}_i$ and $\mathbf{u}_j$ range over all distinct pairs in $\mathcal{V}$, each such pair appearing equally often, so $\mathbf{u}_j - \mathbf{u}_i$ ranges over $\mathcal{V} \setminus \{\mathbf{0}\}$, each non-zero element of $\mathcal{V}$ appearing equally often. It follows that for each $\mathbf{e} \in \mathcal{V}$ we have

$$\overline{f_{\mathbf{e}}(\mathcal{C})} = \frac{|S_\rho(\mathbf{e}) \setminus \{\mathbf{0}\}|}{|\mathcal{V} \setminus \{\mathbf{0}\}|} \le \frac{|S_\rho(\mathbf{e})|}{|\mathcal{V} \setminus \{\mathbf{0}\}|} = \frac{1}{2^n - 1} \sum_{r \le \rho} \binom{n}{r},$$

since there are $\binom{n}{r}$ words at each distance r from $\mathbf{e}$. This upper bound is independent of $\mathbf{e}$, so if we average over all $\mathbf{e}$ then $(C.5)$ gives

$$\overline{\Pr\left(d(\mathbf{u}_j, \mathbf{u}_i + \mathbf{e}) \le \rho\right)} \le \frac{1}{2^n - 1} \sum_{r \le \rho} \binom{n}{r}.$$

This holds for each pair $i \ne j$, so if we sum over all $M(M-1)$ such pairs then $(C.4)$ gives

$$\overline{\Pr_E} < \frac{\delta}{2} + \frac{1}{M} \cdot M(M-1) \cdot \frac{1}{2^n - 1} \sum_{r \le \rho} \binom{n}{r}$$

$$\le \frac{\delta}{2} + \frac{M}{2^n} \sum_{r \le \rho} \binom{n}{r}. \qquad (C.6)$$

Now suppose that η is chosen sufficiently small that

$$Q + \eta < \frac{1}{2}.$$

(possible since $Q < \frac{1}{2}$). Then Exercise 5.7 (with $\lambda = Q + \eta$, so that $\rho = \lambda n$) implies that

$$\sum_{r \leq \rho} \binom{n}{r} = \sum_{r \leq \lambda n} \binom{n}{r} \leq 2^{nH(\lambda)},$$

where H is the binary entropy function H_2, so

$$\overline{\mathrm{Pr}_{\mathrm{E}}} < \frac{\delta}{2} + 2^{-n} M \cdot 2^{nH(\lambda)}$$

$$= \frac{\delta}{2} + M \cdot 2^{n(H(\lambda)-1)}. \tag{C.7}$$

We need an upper bound on the second term in $(C.7)$. Now

$$\log_2\left(M \cdot 2^{n(H(\lambda)-1)}\right) = \log_2 M + n(H(\lambda) - 1)$$
$$= n(R - 1 + H(\lambda)),$$

where $R = \frac{1}{n} \log_2 M$ is the rate of $\mathcal{C}$. Suppose there is some constant $\alpha < 0$ such that

$$R - 1 + H(\lambda) \leq \alpha$$

for all n; then

$$n(R - 1 + H(\lambda)) \leq n\alpha \to -\infty$$

as $n \to \infty$, so $M \cdot 2^{n(H(\lambda)-1)} \to 0$ as $n \to \infty$. Hence

$$M \cdot 2^{n(H(\lambda)-1)} < \frac{\delta}{2}$$

for all sufficiently large n, so $(C.7)$ gives $\overline{\mathrm{Pr}_{\mathrm{E}}} < \delta$. Since the average of Pr_{E} over all $\mathcal{C}$ is less than δ, it follows that at least one code $\mathcal{C}$ must have $\mathrm{Pr}_{\mathrm{E}} < \delta$, as required.

To complete the proof, it remains for us to justify our choices of constants. We need to show that, given $Q < \frac{1}{2}$ and $\varepsilon > 0$, we can find $\eta > 0$ and $\alpha < 0$ (independent of n) such that if n is sufficiently large then

(i) $Q + \eta < \frac{1}{2}$,

(ii) $C - \varepsilon \leq R < C$ (where $C = 1 - H(Q)$ and $R = \frac{1}{n} \log_2 M$ for some $M \in \mathbb{N}$),

(iii) $R - 1 + H(Q + \eta) \leq \alpha$.

Without loss of generality, since $C > 0$ we can take ε sufficiently small that $C - \varepsilon \geq 0$. Now $C = 1 - H(Q)$, the function H is continuous, and $Q < \frac{1}{2}$, so we can choose $\eta > 0$ sufficiently small that both

$$Q + \eta < \frac{1}{2},$$

giving (i), and

$$1 - H(Q + \eta) \geq C - \frac{\varepsilon}{3}.$$

For each $n \geq 3/\varepsilon$ we can choose a rational number $R = k/n$ $(k, n \in \mathbf{N})$ such that

$$C - \varepsilon \leq R \leq C - \frac{2\varepsilon}{3}.$$

We put $M = 2^k$; thus $M \in \mathbf{N}$ and

$$\frac{1}{n} \log_2 M = \frac{k}{n} = R,$$

so (ii) is satisfied. Since $R \leq C - \frac{2\varepsilon}{3}$ and $1 - H(Q + \eta) \geq C - \frac{\varepsilon}{3}$, we have

$$R - 1 + H(Q + \eta) \leq \left(C - \frac{2\varepsilon}{3}\right) - \left(C - \frac{\varepsilon}{3}\right) = -\frac{\varepsilon}{3};$$

this gives (iii) with $\alpha = -\frac{\varepsilon}{3} < 0$, independent of n.   □

Figure C.1 illustrates the relationship between the quantities used here.

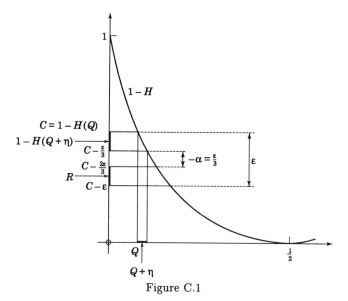

Figure C.1

Solutions to Exercises

Chapter 1

1.1 Use induction on n. If $n = 0$ then $C_n = C$, so $|w| \leq l$. If $n > 0$ then $uw = v$ with $v \in C_{n-1}$ or C, so $|w| \leq |v| \leq l$ by induction or by definition of l respectively. There are only $N = r + r^2 + \cdots + r^l = r(r^l - 1)/(r - 1)$ non-empty r-ary words w with $|w| \leq l$, so $|C_n| \leq N$ for each n. There are only 2^N different sets of such words w, so within the sets $C_0, \ldots, C_{2^N}$ there must be a repetition, $C_j = C_i$ with $i < j \leq 2^N$. By Eq. (1.3), each C_n depends only on C and C_{n-1}, so $C_{j+k} = C_{i+k}$ for all $k \geq 0$; hence each $C_n = C_0$ or C_1 or $\ldots$ or C_{j-1}, so $C_\infty = C_0 \cup C_1 \cup \cdots \cup C_{j-1}$. Thus we have constructed all of C_∞ as soon as we find a repetition among the successive sets $C_0, C_1, \ldots$.

1.2 If $C = \{02, 12, 120, 20, 21\}$ then $C_1 = \{0\}$, $C_1 = \{2\}$, $C_3 = \{0, 1\}$, $C_4 = \{2, 20\}$, $C_5 = \{0, 1\}$; the repetition $C_3 = C_5$ implies that $C_n = \{0, 1\}$ or $\{2, 20\}$ for odd or even $n \geq 3$, so $C_\infty = C_1 \cup \cdots \cup C_4 = \{0, 1, 2, 20\}$. If $C = \{02, 12, 120, 21\}$ then $C_1 = \{0\}$, $C_2 = \{2\}$, $C_3 = \{1\}$, $C_4 = \{2, 20\}$, $C_5 = \{1\}$; again $C_3 = C_5$ implies that $C_n = \{1\}$ or $\{2, 20\}$ for odd or even $n \geq 3$, so $C_\infty = C_1 \cup \cdots \cup C_4 = \{0, 1, 2, 20\}$.

1.3 If $C = \{02, 12, 120, 20, 21\}$ then Exercise 1.2 gives $C_\infty = \{0, 1, 2, 20\}$, containing the code-word 20, so C is not uniquely decodable by Theorem 1.10; for instance 1202120 decodes as 120.21.20 or 12.02.120. If $C = \{02, 12, 120, 21\}$ then Exercise 1.2 gives $C_\infty = \{0, 1, 2, 20\}$, disjoint from C, so C is uniquely decodable.

1.4 Since $u \in C_1$, $u'u = v'$ for some $u', v' \in C$, so $\mathbf{t} = u'uw$ decodes as $u'v$ or $v'w$.

1.5 Since $01,012120 \in C$ we have $2120 \in C_1$; then $212 \in C$ gives $0 \in C_2$, so $01 \in C$ gives $1 \in C_3$, and then $120 \in C$ gives $20 \in C_4$. Thus $20 \in C \cap C_\infty$.

1.6 Since $w \in C_3$ there exist $u \in C, v \in C_2$ with either (i) $uw = v$ or (ii) $vw = u$. Since $v \in C_2$ there exist $u' \in C, v' \in C_1$ with either (a) $u'v = v'$ or (b) $v'v = u'$. Since $v' \in C_1$ there exist $u'', v'' \in C$ with $u''v' = v''$. Now $u, u', u'', v'', w \in C$, so in cases (i)(a), (i)(b), (ii)(a) and (ii)(b) we have the following examples of non-unique decoding: $u''u'uw = u''u'v = u''v' = v''$, $v''uw = u''v'v = u''u'$, $u''u'u = u''u'vw = u''v'w = v''w$, $v''u = u''v'vw = u''u'w$.

1.7 For either code, C_n is non-empty for each $n \geq 1$, so not all infinite code-sequences decode uniquely. For instance $120212121\ldots$ decodes as $120.21.21.\ldots$ or $12.02.12.12.\ldots$.

1.8 $C_1 = \{1, 11\}$ and $C_2 = \{1, 11\}$, so $C_n = \{1, 11\}$ for all $n \geq 1$; thus $C_\infty = \{1, 11\}$, disjoint from C, so C is uniquely decodable by Theorem 1.10. Wait until the sequence of 1s ends; if there are k 1s, where $k \equiv 0, 1$ or $2 \mod (3)$, decode (uniquely) as $0.(111)^{k/3}$, $01.(111)^{(k-1)/3}$ or $011.(111)^{(k-2)/3}$.

1.9 Yes. A first symbol 0 indicates w_1, while a 1 indicates the start of w_2, w_3 or w_4; in the latter case a second symbol 0 indicates w_2, while a 1 indicates w_3 or w_4; in this latter case a third symbol 0 or 1 distinguishes between w_3 and w_4.

1.10 Up to level 2 we have

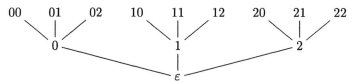

Now attach three vertices $v0, v1, v2$ to each of the nine vertices v at level 2.

1.11 $C = \{0, 10, 110, 111, 2000\}$ is an example. No, since successive choices would eliminate proportions $\frac{1}{3}, \frac{1}{3}, \frac{1}{9}, \frac{1}{9}, \frac{1}{9}, \frac{1}{9}$ of T^*, and these add up to $\frac{10}{9} > 1$.

1.12 No, by Kraft's inequality, since $\sum r^{-l_i} = \frac{28}{27} > 1$. $\{0, 10, 11, 12, 20, 21, 220, 221, 222\}$ is an example. There are 3 choices $(0, 1$ or $2)$ for the code-word of length 1, and then $\binom{6}{5} = 6$ choices for the five code-words of length 2, leaving a unique choice for the three code-words of length 3, so the number of such codes is $3 \times 6 \times 1 = 18$.

1.13 If $j \geq 2$ then $\mathbf{t} = \mathbf{t}'w$ with last code-word (well-defined since $\mathcal{C}$ is uniquely decodable) $w = 0, 10$ or 11. If $w = 0$, there are N_{j-1} possibilities for $\mathbf{t}'$ (of length $j-1$); if $w = 10$ or 11, there are N_{j-2} possibilities for $\mathbf{t}'$ (of length $j-2$) in each case. Hence $N_j = N_{j-1} + 2N_{j-2}$. This 2nd-order linear recurrence relation has auxiliary equation $\lambda^2 = \lambda + 2$, with roots $\lambda = 2, -1$, so the general solution is $N_j = A.2^j + B.(-1)^j$. The initial conditions $N_1 = 1$ ($\mathbf{t} = 0$) and $N_2 = 3$ ($\mathbf{t} = 00, 10, 11$) give $A = 2/3, B = 1/3$, so $N_j = (2^{j+1} + (-1)^j)/3$. (See Chapter 4 of [An74] for recurrence relations.)

1.14 In the proof of Theorem 1.20, there are r^{l_1} choices for w_1, then (after pruning) $r^{l_2} - r^{l_2 - l_1} = r^{l_2}(1 - r^{-l_1})$ choices for w_2, then $r^{l_3} - r^{l_3 - l_1} - r^{l_3 - l_2} = r^{l_3}(1 - r^{-l_1} - r^{-l_2})$ choices for w_3, etc., giving $r^{l_1 + l_2 + \cdots + l_q}(1 - r^{-l_1}) \ldots (1 - r^{-l_1} - \cdots - r^{-l_{q-1}})$ choices for $w_1, \ldots, w_q$.

1.15 $\mathcal{C}$ is exhaustive if and only if every leaf of $T^{\leq l}$ is above a code-word. The codes in Examples 1.16 and 1.18 are exhaustive.

1.16 Imitate the proof of Theorem 1.20: $\mathcal{C}$ is exhaustive if and only if every leaf of $T^{\leq l}$ lies above a code-word; there are r^l leaves, and each code-word of length l_i is below $r^{l - l_i}$ leaves, so this implies $r^l \leq \sum_i r^{l - l_i}$, that is, $\sum_i r^{-l_i} \geq 1$. Equality occurs here if and only if each leaf lies above a unique code-word, that is, $\mathcal{C}$ is a prefix code, or equivalently, instantaneous.

1.17 By Exercise 1.16, if (b) is true then (a) is equivalent to (c); thus (a) and (b) imply (c), and (b) and (c) imply (a). If (a) and (c) are true, then in the proof of Theorem 1.20 every leaf of $T^{\leq l}$ is above a code-word, giving (b). If $T = \mathbf{Z}_2$ then the codes $\{0\}, \{0, 1, 00\}$ and $\{0, 00, 01\}$ satisfy (a), (b) and (c) alone, so none of (a), (b) or (c) implies any other.

Chapter 2

2.1 Let $p_i > p_j$ with $l_i > l_j$. Transposing the code-words w_i and w_j in $\mathcal{C}$ gives another instantaneous code $\mathcal{C}^*$. The summands $p_i l_i$ and $p_j l_j$ in $L(\mathcal{C})$ are replaced with $p_i l_j$ and $p_j l_i$ in $L(\mathcal{C}^*)$. Then $(p_i l_i + p_j l_j) - (p_i l_j + p_j l_i) = (p_i - p_j)(l_i - l_j) > 0$ gives $L(\mathcal{C}) > L(\mathcal{C}^*)$, contradicting the optimality of $\mathcal{C}$. Hence $l_i \leq l_j$.

2.2 S determines a vector $\mathbf{p} = (p_1, \ldots, p_q) \in \mathbf{R}^q$ with $p_i \geq 0$ and $\sum p_i = 1$, and each code $\mathcal{C}$ determines a vector $\mathbf{l} = (l_1, \ldots, l_q) \in \mathbf{N}^q \subset \mathbf{R}^q$, so that $L(\mathcal{C}) = \sum p_i l_i = \mathbf{p}.\mathbf{l}$. Given $\mathbf{p}$, the problem is to show that some instantaneous code minimises $\mathbf{p}.\mathbf{l}$. The proof of Theorem 2.3 shows that, since each $l_i \in$

N, there are only finitely many possible values of **p**.l not exceeding some constant; among the finitely many corresponding to instantaneous codes, one can choose a least value, corresponding to an optimal code.

2.3 One solution is $\mathcal{C} = \{0, 10, 1100, 1101, 1110, 1111\}$, with $L(\mathcal{C}) = \sum_i p_i l_i = 2.2$. Another possibility is $\mathcal{C} = \{1, 00, 011, 0100, 01010, 01011\}$, so $\mathcal{C}$ and $\{l_i\}$ are not unique. However, $L(\mathcal{C})$ is unique by the optimality of Huffman codes.

2.4 When $q = 3$, $\mathcal{C} = \{0, 10, 11\}$ has $L(\mathcal{C}) = p_1 + 2p_2 + 2p_3 = 2 - p_1$ (since $\sum p_i = 1$). When $q = 4$, $l_i = 1, 2, 3, 3$ or $2, 2, 2, 2$ as $p_3 + p_4 \leq p_1$ or $p_3 + p_4 \geq p_1$, giving $L(\mathcal{C}) = p_1 + 2p_2 + 3p_3 + 3p_4 = 3 - 2p_1 - p_2$ or $2p_1 + 2p_2 + 2p_3 + 2p_4 = 2$ respectively.

2.5 In Exercise 2.3, the merged probabilities $p', p'', \ldots$ are $0.1, 0.2, 0.3, 0.6, 1$ with sum 2.2. In Exercise 2.4, with $q = 3$, they are $p' = p_2 + p_3$ and $p'' = 1$, with $p' + p'' = p_2 + p_3 + 1 = 2 - p_1$; when $q = 4$ they are $p' = p_3 + p_4$, then $p'' = p_2 + p_3 + p_4$ or $p_1 + p_2$ as $p_3 + p_4 \leq p_1$ or $p_3 + p_4 \geq p_1$, and $p''' = 1$, with $p' + p'' + p''' = 1 + p_2 + 2p_3 + 2p_4 = 3 - 2p_1 - p_2$ or 2 respectively.

2.6 The proof that Huffman codes are optimal is by comparing Huffman codes with optimal codes. This assumes that every source has an optimal code, so the argument is circular.

2.7 Binary: $\mathcal{C} = \{00, 10, 010, 110, 111, 0110, 01110, 01111\}$ with $L(\mathcal{C}) = 2.72$. Ternary: $\mathcal{C} = \{0, 10, 11, 12, 20, 21, 220, 221\}$ with $L(\mathcal{C}) = 1.77$.

2.8 The argument is the same as in §2.3, except that $s_{q-r+1}, \ldots, s_q$ are amalgamated, with $L(\mathcal{C}) - L(\mathcal{C}') = p_{q-r+1}(l + 1) + \cdots + p_q(l + 1) - (p_{q-r+1} + \cdots + p_q)l = p_{q-r+1} + \cdots + p_q = p'$. Any extra symbols s_i adjoined have no effect on $L(\mathcal{C})$, since $p_i = 0$.

2.9 $\mathcal{S}^3$ has $2^3 = 8$ symbols with probabilities $8/27, 4/27, 4/27, 4/27, 2/27, 2/27, 2/27, 1/27$. In Huffman coding, the merged probabilities are $3/27, 4/27, 7/27, 8/27, 11/27, 16/27, 1$, with sum $L_3 = 76/27$.

2.10 There are 24 optimal binary codes, the 4! permutations of $\{00, 01, 10, 11\}$ (these have $L(\mathcal{C}) = 2$, whereas any other instantaneous code has $L(\mathcal{C}) \geq 2.1$). Of these, eight are Huffman codes, namely those in which the last two code-words (those with lowest probabilities) are siblings: in constructing the codes $\mathcal{C}'', \mathcal{C}'$ and $\mathcal{C}$ there are two possibilities ($w'0, w'1$ or $w'1, w'0$) at each of the three stages, giving $2^3 = 8$ possibilities for $\mathcal{C}$.

2.11 The given inequalities imply that $s' = s_{q-1} \vee s_q$, and then $s^{(k)} = s_{q-k} \vee s^{(k-1)} = s_{q-k} \vee \cdots \vee s_q$ for $1 < k \leq q - 1$, using induction on k. Thus for $i \leq q - 1$, s_i is amalgamated i times (in $\mathcal{S}^{(q-1-i)}, \ldots, \mathcal{S}^{(q-2)}$), giving $l_i = i$, while s_q is amalgamated $q - 1$ times, giving $l_q = q - 1$. In assigning code-words there are just two choices at each of the $q-1$ stages: a code-word $w^{(k)}$ for $s^{(k)}$ generates code-words $w^{(k)}0$ and $w^{(k)}1$ for s_{q-k} and $s^{(k-1)}$ in either order. Hence there are 2^{q-1} binary Huffman codes for S. The probabilities $p_i = 2^{-i}$ for $i = 1, \ldots, q - 1$ and $p_q = 2^{1-q}$ satisfy the given conditions, since $p_{i+2} + \cdots + p_q = 2^{-i-1} < p_i$ for $i = 1, \ldots, q - 3$.

2.12 In r-ary Huffman coding, a code-word $w' \in \mathcal{C}'$ of length l is replaced with r code-words of length $l+1$ in $\mathcal{C}$, so $\sigma(\mathcal{C}) - \sigma(\mathcal{C}') = r(l+1) - l = (r-1)l + r$. Since r is fixed, one can minimise $\sigma(\mathcal{C})$ by minimising l at each stage of the algorithm. This is achieved by placing each amalgamated symbol s' as early as possible among the ordered symbols of $\mathcal{S}'$, whenever its probability coincides with others. In the given example, $s' = s_3 \vee s_4$ has probability $p' = 1/3$; since $p_1 = p_2 = 1/3$ also, there are three possible places for s' in $\mathcal{S}'$; placing it before p_1 ensures that $l = 1$ rather than 2, and this yields $\sigma(\mathcal{C}) = 2 + 2 + 2 + 2 = 8$ rather than $\sigma(\mathcal{C}) = 1 + 2 + 3 + 3 = 9$.

2.13 Construct a binary Huffman code $\mathcal{C} = \{w_1, \ldots, w_q\}$ using the probabilities p_i for the objects s_i. For each k, let T_k be the set of objects whose code-word has 1 in position k, and let Q_k be the question "Is s in T_k?". Then asking $Q_1, Q_2, \ldots$ identifies w_i and hence s_i after $l_i = |w_i|$ questions, so the average number of questions needed is $\sum_i p_i l_i = L(\mathcal{C})$. Similarly any other sequence of questions would, by assigning symbols 1 or 0 to successive answers "yes" or "no", correspond to another binary prefix code for S, so the Huffman code, being optimal, corresponds to a best possible sequence.

2.14 Yes: if $q = 3$ then $L(\mathcal{C}) = 5/3$; $\mathcal{S}^2$ has nine equiprobable symbols, giving $L_2 = 29/9$ by (2.4), so $L_2/2 = 29/18 < L(\mathcal{C})$. If $q = 2^l$ for some integer l, then in $\mathcal{C}$ each $l_i = l$ and hence $L(\mathcal{C}) = l$; similarly $\mathcal{S}^n$ has 2^{ln} equiprobable symbols, so $L_n = ln$ and hence $L_n/n = L(\mathcal{C})$ for all n.

Chapter 3

3.1 $H_2(\mathcal{S}) = \sum_i p_i \log_2(1/p_i) \approx 2.681$ and $H_3(\mathcal{S}) = \sum_i p_i \log_3(1/p_i) \approx 1.691$. Binary and ternary Huffman codes have average word-lengths $L(\mathcal{C}) = 2.92$ and 1.77 respectively.

3.2 Let S have probabilities $p_i = 2^{-1}, 2^{-2}, 2^{-3}, \ldots, 2^{2-q}, 2^{1-q}, 2^{1-q}$; let C have code-words $0, 10, 110, \ldots, 1 \ldots 10, 1 \ldots 10, 1 \ldots 11$ of lengths $1, 2, 3, \ldots, q-2, q-1, q-1$. Then $H_2(S) = 2^{-1}.1 + 2^{-2}.2 + 2^{-3}.3 + \cdots + 2^{2-q}.(q-2) + 2.2^{1-q}.(q-1) = L(C)$.

3.3 $H_2(S) = \sum_i p_i \log_2(1/p_i) \approx 2.144$. A Huffman code C has $L(C) = 2.2$ by Exercise 2.3, so $\eta = H/L \approx 2.144/2.2 \approx 0.975$.

3.4 A Shannon–Fano code has $l_i = \lceil \log_2(1/p_i) \rceil = 2, 2, 4, 4, 5, 5$, so $L = 2.7$ and $\eta \approx 2.144/2.7 \approx 0.794$.

3.5 $H_2(S) = -\frac{4}{5}\log_2\frac{4}{5} - \frac{1}{5}\log_2\frac{1}{5} = \log_2 5 - \frac{8}{5}$. The extention S^n has $\binom{n}{k}$ symbols of probability $(4/5)^k(1/5)^{n-k} = 4^k/5^n$ for each $k = 0, \ldots, n$, each given a code-word of length $\lceil \log_2(5^n/4^k) \rceil = \lceil n \log_2 5 \rceil - 2k$, so if a_n denotes $\lceil n \log_2 5 \rceil$ then a binary Shannon–Fano code for S^n has average word-length

$$L_n = \sum_{k=0}^{n} \binom{n}{k} \frac{4^k}{5^n}(a_n - 2k) = \frac{1}{5^n}\left(a_n \sum_{k=0}^{n} \binom{n}{k}4^k - 2 \sum_{k=0}^{n} k\binom{n}{k}4^k \right) = a_n - \frac{8n}{5},$$

as in §3.7. As $n \to \infty$, $a_n/n \to \log_2 5$, so $\frac{1}{n}L_n \to \log_2 5 - \frac{8}{5} = H_2(S)$.

3.6 S^n has q^n symbols, all of probability $1/q^n$, so each is given an r-ary code-word of length $\lceil \log_r q^n \rceil = \lceil n \log_r q \rceil$. Thus $L_n = \lceil n \log_r q \rceil$ and hence $\frac{1}{n}L_n = \lceil n \log_r q \rceil / n \to \log_r q = H_r(S)$ as $n \to \infty$.

3.7 Define $g(x) = f(e^{-x})$, a strictly increasing function on $[0, +\infty)$ with $g(x + y) = g(x) + g(y)$ for all $x, y \geq 0$ (which extends to all finite sums). Putting $x = y = 0$ shows that $g(0) = 0$. Define $c = g(1)$, so $c > g(0) = 0$. We will show that $g(x) = cx$ for all $x \geq 0$, so $f(x) = -c \ln x = -\log_r x$ as required, with $r = e^{1/c} > 0$. Induction on n gives $g(2^n) = c2^n$ for all integers $n \geq 0$. Also $c = g(1) = g(\frac{1}{2}) + g(\frac{1}{2})$, so $g(\frac{1}{2}) = c/2$, and induction gives $g(2^n) = c2^n$ for all $n < 0$. Each $x \geq 0$ has a binary expansion $x = \sum_{n=-\infty}^{N} a_n 2^n$ with each $a_n = 0$ or 1, so $\sum_{n=M}^{N} a_n 2^n \leq x \leq \sum_{n=M}^{N} a_n 2^n + 2^M$ for each $M \leq N$. Applying g to these inequalities and using its additive and increasing properties gives $\sum_{n=M}^{N} c a_n 2^n \leq g(x) \leq \sum_{n=M}^{N} c a_n 2^n + c2^M$. Dividing by c and then letting $M \to -\infty$ we see that $g(x)/c = x$, as required.

3.8 $s_i = 2, 3, \ldots, 12$ with probabilities $p_i = \frac{1}{36}, \frac{2}{36}, \frac{3}{36}, \frac{4}{36}, \frac{5}{36}, \frac{6}{36}, \frac{5}{36}, \frac{4}{36}, \frac{3}{36}, \frac{2}{36}, \frac{1}{36}$, giving $H_2(S) = \sum_i p_i \log_2(1/p_i) = (23 + 30\log_2 3 - 5\log_2 5)/18 \approx 3.2745$. In Huffman coding the successive merged probabilities are $p' = \frac{2}{36}, \frac{4}{36}, \frac{5}{36}, \frac{7}{36}, \frac{8}{36}, \frac{10}{36}, \frac{11}{36}, \frac{15}{36}, \frac{21}{36}, 1$, with sum $L(C) = 119/36 = 3.30555\ldots$.

In Shannon–Fano coding $l_i = \lceil \log_2(1/p_i) \rceil = 6, 5, 4, 4, 3, 3, 3, 4, 4, 5, 6$ so $L(\mathcal{C}) = \sum_i p_i l_i = 136/36 = 3.777\ldots$.

3.9

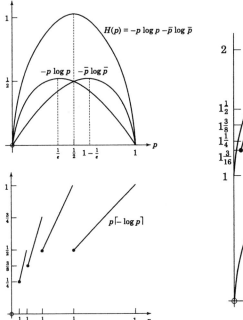

The graph of $\bar{p}\lceil - \log \bar{p}\,\rceil$ is the mirror-image

3.10 By Corollary 3.12, $L(\mathcal{C}) = H_r(\mathcal{S})$ if and only if $\mathcal{S}$ has probabilities $p_i = r^{e_i}$ for integers $e_1, \ldots, e_q \le 0$. In this case, $\sum_{i=1}^{q} r^{e_i} = \sum_{i=1}^{q} p_i = 1$, so if $e = \min e_i$ then $\sum_{i=1}^{q} r^{e_i - e} = r^{-e}$ with $e_i - e, -e \ge 0$; each term $r^{e_i - e}, r^{-e} \equiv 1$ mod $(r-1)$, so $q \equiv 1$ mod $(r-1)$. Conversely, if $q = 1 + k(r-1)$, let $\mathcal{S}$ have $r - 1$ symbols of probability r^{-l} for each $l = 1, \ldots, k-1$, and r of probability r^{-k}, so $\sum_{i=1}^{q} p_i = (r-1) \sum_{l=1}^{k-1} r^{-l} + r.r^{-k} = 1$; then Corollary 3.12 gives $L(\mathcal{C}) = H_r(\mathcal{S})$.

3.11 $H_3(\mathcal{S}) = -\frac{3}{4}\log_3 \frac{3}{4} - \frac{1}{4}\log_3 \frac{1}{4} = \log_3 4 - \frac{3}{4}$. The extention $\mathcal{S}^n$ has $\binom{n}{k}$ symbols of probability $(3/4)^k (1/4)^{n-k} = 3^k/4^n$ for each $k = 0, \ldots, n$, each given a code-word of length $\lceil \log_3(4^n/3^k) \rceil = \lceil n \log_3 4 \rceil - k$, so if a_n denotes $\lceil n \log_3 4 \rceil$ then a ternary Shannon–Fano code for $\mathcal{S}^n$ has average word-length

$$L_n = \sum_{k=0}^{n} \binom{n}{k} \frac{3^k}{4^n}(a_n - k) = \frac{1}{4^n}\left(a_n \sum_{k=0}^{n} \binom{n}{k}3^k - \sum_{k=0}^{n} k\binom{n}{k}3^k\right) = a_n - \frac{3n}{4},$$

as in §3.7. As $n \to \infty$, $a_n/n \to \log_3 4$, so $\frac{1}{n}L_n \to \log_3 4 - \frac{3}{4} = H_3(\mathcal{S})$. In binary Shannon–Fano coding, a symbol of probability $3^k/4^n$ gets a code-word of length $\lceil \log_2(4^n/3^k) \rceil = 2n - b_k$ where $b_k = \lfloor k \log_2 3 \rfloor$, so

$$L_n = \sum_{k=0}^{n} \binom{n}{k} \frac{3^k}{4^n}(2n - b_k) = 2n - \frac{1}{4^n}\sum_{k=0}^{n}\binom{n}{k}3^k b_k.$$

There is no simple way of evaluating this last sum; however $k \log_2 3 - 1 < b_k \le k \log_2 3$, so

$$\frac{1}{4^n}\sum_{k=0}^{n}\binom{n}{k}3^k(k\log_2 3 - 1) < \frac{1}{4^n}\sum_{k=0}^{n}\binom{n}{k}3^k b_k \le \frac{1}{4^n}\sum_{k=0}^{n}\binom{n}{k}3^k k\log_2 3$$

giving

$$\frac{3n}{4}\log_2 3 - 1 < \frac{1}{4^n}\sum_{k=0}^{n}\binom{n}{k}3^k b_k \le \frac{3n}{4}\log_2 3,$$

and hence $\frac{1}{n}L_n \to 2 - \frac{3}{4}\log_2 3 = H_2(\mathcal{S})$ as $n \to \infty$.

3.12 Let $p_1 = 1 - \delta$ and $p_2 = \cdots = p_q = \delta/(q-1)$, where $0 < \delta < 1$. Then $H_r(\mathcal{S}) = -(1-\delta)\log_r(1-\delta) - \delta\log_r(\delta/(q-1)) \to 0$ as $\delta \to 0$ (since $x \log_r x \to 0$ as $x \to 0$ or $x \to 1$), so $H_r(\mathcal{S}) < \varepsilon$ for sufficiently small δ. Every instantaneous code $\mathcal{C}$ has $L(\mathcal{C}) \ge 1$, so $L(\mathcal{C}) > 1 + H_r(\mathcal{S}) - \varepsilon$.

3.13 Define $H_r(\mathcal{S}) = \sum_{k=1}^{\infty} p_k \log_r(1/p_k) = -\sum_{k=1}^{\infty} p_k \log_r p_k$. If $p_k = 2^{-k}$ then $H_2(\mathcal{S}) = \sum_{k=1}^{\infty} 2^{-k}\log_2 2^k = \sum_{k=1}^{\infty} 2^{-k}k = 2$. (For the last step, differentiate $(1-x)^{-1} = 1 + x + x^2 + \cdots$, multiply by x, then put $x = \frac{1}{2}$.) The prefix code $\mathcal{C} = \{0, 10, 110, 1110, \ldots\}$ is instantaneous, and $L(\mathcal{C}) = \sum_{k=1}^{\infty} 2^{-k}k = 2 = H_2(\mathcal{S})$.

3.14 If $X_n = s_i$, the uncertainty about X_{n+1} is the conditional entropy $H(\mathcal{S} \mid X_n = s_i) = -\sum_j p_{ij}\log p_{ij}$; averaging over s_i gives $\sum_i p_i(-\sum_j p_{ij}\log p_{ij})$ $= -\sum_i\sum_j p_i p_{ij}\log p_{ij}$ as the average uncertainty about $\mathcal{S}$. The numbers $p_i p_{ij} = \Pr(X_n = s_i, X_{n+1} = s_j)$ form a probability distribution (for the extension $\mathcal{S}^2$), as do the numbers $p_i p_j$ (for $\mathcal{T}^2$), so Corollary 3.9 gives $-\sum_i\sum_j p_i p_{ij}\log(p_i p_{ij}) \le -\sum_i\sum_j p_i p_{ij}\log(p_i p_j)$ and hence (by the additivity of logarithms) $-\sum_i\sum_j p_i p_{ij}\log p_{ij} \le -\sum_i\sum_j p_i p_{ij}\log p_j$. Since $\sum_i p_i p_{ij} = p_j$, this gives $H(\mathcal{S}) \le H(\mathcal{T})$. Corollary 3.9 gives equality if and only if $p_i p_{ij} = p_i p_j$ for all i, j, that is, X_{n+1} and X_n are statistically independent. The interpretation is that knowing the probabilities p_{ij} generally decreases our uncertainty about $\mathcal{S}$. Since $\sum_i p_i p_{ij} = p_j$, (p_i) is an eigenvector of the matrix (p_{ij}) with eigenvalue $\lambda = 1$, satisfying $\sum_i p_i = 1$. In this case $p_1 = p_3 = 1/4$ and $p_2 = 1/2$, so $H(\mathcal{T}) = 3/2$ and $H(\mathcal{S}) = (2 + 9\log 3)/12 \approx 1.355$.

Chapter 4

4.1 If Γ has input symbols a_i and output symbols b_j, and if Γ' has input symbols b_j and output symbols c_k, then $\Pr(c_k|a_i) = \sum_j \Pr(b_j|a_i)\Pr(c_k|b_j)$. This is the rule for matrix multiplication, so if M and M' are the channel matrices for Γ and Γ', then the composite channel $\Gamma \circ \Gamma'$ has channel matrix MM'. More generally, if channels $\Gamma_1, \ldots, \Gamma_n$ have matrices $M_1, \ldots, M_n$, and the output of Γ_i is the input of Γ_{i+1} for $i = 1, \ldots, n-1$, then induction on n shows that $\Gamma_1 \circ \cdots \circ \Gamma_n$ has channel matrix $M_1 \cdots M_n$.

4.2 (i) $Q_{00} = pP/q$ and $Q_{10} = \bar{p}\bar{P}/q$, so $Q_{00} < Q_{10}$ if and only if $pP < \bar{p}\bar{P}$; similarly, $Q_{01} = p\bar{P}/\bar{q}$ and $Q_{11} = \bar{p}P/\bar{q}$, so $Q_{01} < Q_{11}$ if and only if $p\bar{P} < \bar{p}P$. Equivalently, $p = p(P + \bar{P}) < (\bar{p} + p)\bar{P} = \bar{P}$ and $p = p(P + \bar{P}) < (p + \bar{p})P = P$, that is, $p < \min(P, \bar{P})$. Whether 0 or 1 is received, it is most likely that 1 was transmitted.

(ii) $pP > \bar{p}\bar{P}$ and $p\bar{P} < \bar{p}P$, or equivalently $\bar{P} < p < P$. If 0 or 1 is received, that symbol is most likely to have been transmitted.

(iii) $pP < \bar{p}\bar{P}$ and $p\bar{P} > \bar{p}P$, or equivalently $P < p < \bar{P}$. If 0 or 1 is received, the other symbol is most likely to have been transmitted.

4.3 Using $R_{ij} = q_j Q_{ij}$ and $\sum_i R_{ij} = q_j$ we have

$$H(\mathcal{A}, \mathcal{B}) = \sum_i \sum_j R_{ij} \log \frac{1}{R_{ij}} = \sum_i \sum_j R_{ij} \log \frac{1}{q_j} + \sum_i \sum_j R_{ij} \log \frac{1}{Q_{ij}}$$

$$= \sum_j q_j \log \frac{1}{q_j} + \sum_i \sum_j R_{ij} \log \frac{1}{Q_{ij}} = H(\mathcal{B}) + H(\mathcal{A} \mid \mathcal{B}).$$

Thus $H(\mathcal{A} \mid \mathcal{B}) = H(\mathcal{A}, \mathcal{B}) - H(\mathcal{B})$ is the information gained by the receiver (who already knows $\mathcal{B}$) if he discovers $\mathcal{A}$. Equivalently, it is his uncertainty about $\mathcal{A}$, given $\mathcal{B}$.

4.4 By Lemma 3.21, if $\mathcal{S}$ and $\mathcal{T}$ are independent sources then $H(\mathcal{S} \times \mathcal{T}) = H(\mathcal{S}) + H(\mathcal{T})$. If Γ and Γ' have inputs $\mathcal{A}, \mathcal{A}'$ and outputs $\mathcal{B}, \mathcal{B}'$, this immediately gives $H(\mathcal{A} \times \mathcal{A}') = H(\mathcal{A}) + H(\mathcal{A}')$, and similarly for $H(\mathcal{B} \times \mathcal{B}')$ and $H(\mathcal{A} \times \mathcal{A}', \mathcal{B} \times \mathcal{B}')$. If b_j and b'_k are typical output symbols of Γ and

Γ', with probabilities q_j and q'_k, then

$$H(\mathcal{A} \times \mathcal{A}' \mid \mathcal{B} \times \mathcal{B}') = \sum_{j,k} q_j q'_k H(\mathcal{A} \times \mathcal{A}' \mid b_j b'_k)$$

$$= \sum_{j,k} q_j q'_k \left(H(\mathcal{A} \mid b_j) + H(\mathcal{A}' \mid b'_k) \right)$$

$$= \sum_j q_j H(\mathcal{A} \mid b_j) + \sum_k q'_k H(\mathcal{A}' \mid b'_k)$$

$$= H(\mathcal{A} \mid \mathcal{B}) + H(\mathcal{A}' \mid \mathcal{B}'),$$

using $\sum_j q_j = \sum_k q'_k = 1$ in the penultimate line. A similar argument shows that $H(\mathcal{B} \times \mathcal{B}' \mid \mathcal{A} \times \mathcal{A}') = H(\mathcal{B} \mid \mathcal{A}) + H(\mathcal{B}' \mid \mathcal{A}')$. The corresponding results for Γ^n follow by induction on n.

4.5 Suppose the result is false, so $f(c) \leq \lambda f(a) + \overline{\lambda} f(b)$ for some $c = \lambda a + \overline{\lambda} b$ where $a < b$ and $0 < \lambda < 1$ (so $a < c < b$). The Mean Value Theorem (applied to f on $[a, c]$ and $[c, b]$) gives $f'(c_1) = (f(c) - f(a))/(c - a)$ and $f'(c_2) = (f(b) - f(c))/(b - c)$ for some c_1, c_2 where $a < c_1 < c < c_2 < b$. Substituting for c and using the inequality for $f(c)$ gives

$$f'(c_1) \leq \frac{\lambda f(a) + \overline{\lambda} f(b) - f(a)}{\lambda a + \overline{\lambda} b - a} = \frac{f(b) - f(a)}{b - a}$$

$$= \frac{f(b) - \lambda f(a) - \overline{\lambda} f(b)}{b - \lambda a - \overline{\lambda} b} \leq f'(c_2),$$

so the Mean Value Theorem (applied to f' on $[c_1, c_2]$) gives $f''(c_3) = (f'(c_2) - f'(c_1))/(c_2 - c_1) \geq 0$ for some c_3 where $c_1 < c_3 < c_2$. This contradicts $f''(x) < 0$ for all $x \in (0, 1)$.

4.6 Let Γ be a binary channel with matrix $\left(\begin{smallmatrix} 1 & 0 \\ 1 & 0 \end{smallmatrix} \right)$, so every input symbol $a = 0$ or 1 is transmitted as $b = 0$. If the input probabilities are $p, \overline{p}$ then $H(\mathcal{A}) = H(p)$, while $H(\mathcal{B}) = H(1) = 0$ since the output probabilities are $1, 0$. Thus $H(\mathcal{A}) > H(\mathcal{B})$ if $0 < p < 1$.

4.7 The input symbols 0 and 1 have probabilities p and $\overline{p}$, so $H(\mathcal{A}) = H(p)$. The output symbols $0, 1$ and ? have probabilities $pP, \overline{p}P$ and $\overline{P}$, so

$$H(\mathcal{B}) = -pP \log pP - \overline{p}P \log \overline{p}P - \overline{P} \log \overline{P}$$

$$= -P(p \log p + \overline{p} \log \overline{p} + \log P) - \overline{P} \log \overline{P} = PH(p) + H(P),$$

$$H(\mathcal{B} \mid \mathcal{A}) = -pP \log P - p\overline{P} \log \overline{P} - \overline{p}P \log P - \overline{p}\,\overline{P} \log \overline{P}$$

$$= -P \log P - \overline{P} \log \overline{P} = H(P),$$

$$H(\mathcal{A}, \mathcal{B}) = H(\mathcal{A}) + H(\mathcal{B} \mid \mathcal{A}) = H(p) + H(P) \quad \text{by (4.6)},$$

$$H(\mathcal{A} \mid \mathcal{B}) = H(\mathcal{A}, \mathcal{B}) - H(\mathcal{B}) = H(p) - PH(p) = \overline{P}H(p) \quad \text{by (4.7)}.$$

Then $P, H(p) \geq 0$ gives $H(\mathcal{B} \mid \mathcal{A}) \leq H(\mathcal{B})$, and $\overline{P} \leq 1$ gives $H(\mathcal{A} \mid \mathcal{B}) \leq H(\mathcal{A})$.

4.8

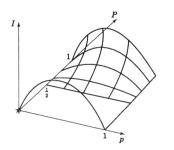 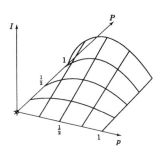

4.9 By Exercise 4.7, the BEC has $I(\mathcal{A}, \mathcal{B}) = H(\mathcal{B}) - H(\mathcal{B} \mid \mathcal{A}) = PH(p)$. With P fixed and p varying, $I(\mathcal{A}, \mathcal{B})$ is maximised when $p = 1/2$ (so $H(p) = 1$), giving $C = I_{\max} = P$.

4.10 If Γ and Γ' have inputs $\mathcal{A}, \mathcal{A}'$ and outputs $\mathcal{B}, \mathcal{B}'$, then Exercise 4.4 gives $H(\mathcal{B} \times \mathcal{B}') = H(\mathcal{B}) + H(\mathcal{B}')$ and $H(\mathcal{B} \times \mathcal{B}' \mid \mathcal{A} \times \mathcal{A}') = H(\mathcal{B} \mid \mathcal{A}) + H(\mathcal{B}' \mid \mathcal{A}')$. Subtracting gives $I(\mathcal{A} \times \mathcal{A}', \mathcal{B} \times \mathcal{B}') = I(\mathcal{A}, \mathcal{B}) + I(\mathcal{A}', \mathcal{B}')$, and taking maxima over all $\mathcal{A}$ and $\mathcal{A}'$ shows that $\Gamma \times \Gamma'$ has capacity $C + C'$. It follows by induction on n that Γ^n has capacity nC.

4.11 If $\mathbf{p} = (p_i) \in \mathcal{P}$ then $0 \leq p_i \leq 1$ for $i = 1, \dots, r$, so $|\mathbf{p}|^2 = \sum_i p_i^2 \leq r$; thus $\mathcal{P}$ is bounded. To show that $\mathcal{P}$ is closed, let $\mathbf{y} = (y_i) \in \mathbf{R}^r \setminus \mathcal{P}$, so either some $y_i < 0$ or $\sum y_i \neq 1$. In the first case, all $\mathbf{x} \in \mathbf{R}^r$ with $|\mathbf{x} - \mathbf{y}| < |y_i|$ satisfy $x_i < 0$, since $|x_i - y_i| \leq |\mathbf{x} - \mathbf{y}|$, so $\mathbf{x} \notin \mathcal{P}$. In the second case, $\mathbf{y}$ is distance $d = |\sum y_i - 1|/\sqrt{r} > 0$ from the hyperplane $\sum p_i = 1$, so all $\mathbf{x}$ with $|\mathbf{x} - \mathbf{y}| < d$ are outside $\mathcal{P}$.

4.12 Γ has channel matrix $M = \begin{pmatrix} P & Q \\ Q & P \end{pmatrix}$, where $Q = \overline{P} = 1 - P$, so Γ^n has channel matrix M^n by Exercise 4.1. By induction on n, M^n has the form $\begin{pmatrix} P_n & Q_n \\ Q_n & P_n \end{pmatrix}$ where $0 \leq P_n \leq 1$ and $Q_n = \overline{P_n}$, so Γ^n is a BSC. Now M has eigenvalues $\lambda = 1, 2P - 1$, so M^n has eigenvalues $\lambda^n = 1, (2P - 1)^n$; thus $2P_n = \operatorname{tr}(M^n) = 1 + (2P - 1)^n$, giving $P_n = (1 + (2P - 1)^n)/2$, $Q_n = (1 - (2P - 1)^n)/2$ (alternatively, prove these by induction on n). Thus Γ^n has capacity $C_n = 1 - H(P_n) = 1 - H((1 + (2P - 1)^n)/2)$. As $n \to \infty$, $(2P - 1)^n \to 0$ (provided $0 < P < 1$), so $C_n \to 1 - H(\frac{1}{2}) = 0$. If $P = 0$ or 1 then each $P_n = 0$ or 1 also, so $C_n = 1$ for all n.

4.13 $C = I_{\max}$, so $C = 0$ if and only if $I(\mathcal{A}, \mathcal{B}) = 0$ for all $\mathcal{A}$, i.e. (by Theorem 4.11) $\mathcal{A}$ and $\mathcal{B}$ are independent for all $\mathcal{A}$. This means that

$P_{ij} = \Pr(b_j \mid a_i) = \Pr(b_j)$ for all i and j, i.e. the rows of M are all equal. The interpretation is that the input probability distribution has no effect on the output distribution, so the receiver gains no information about the input.

4.14 Multiplying H by a constant if necessary, we can take $r = e$, so $H(\mathbf{x}) = -\sum_i x_i \ln x_i$ and hence $\partial H/\partial x_i = -1 - \ln x_i$ for $x_i > 0$. If $\mathbf{p} \neq \mathbf{q}$ in $\mathcal{P}$ then the function $f(\lambda) = H(\lambda\mathbf{p} + \overline{\lambda}\mathbf{q})$ is continuous on $[0,1]$, with $f'(\lambda) = -\sum_i(1 + \ln(\lambda p_i + \overline{\lambda}q_i))(p_i - q_i)$ and $f''(\lambda) = -\sum_i(p_i - q_i)^2/(\lambda p_i + \overline{\lambda}q_i)$ for all $\lambda \in (0,1)$, where we sum over all i with $\lambda p_i + \overline{\lambda}q_i > 0$. Thus $f''(\lambda) < 0$ on $(0,1)$, so f is strictly convex on $[0,1]$ by Lemma 4.6. Hence $H(\lambda\mathbf{p} + \overline{\lambda}\mathbf{q}) \geq \lambda H(\mathbf{p}) + \overline{\lambda}H(\mathbf{q})$ for all $\lambda \in [0,1]$, with equality if and only if $\lambda = 0$ or 1.

4.15 $I(\mathcal{A},\mathcal{B}) = H(\mathcal{B}) - H(\mathcal{B} \mid \mathcal{A})$ with $H(\mathcal{B} \mid \mathcal{A}) = -\sum_i p_i(\sum_j P_{ij} \log P_{ij})$. The condition on rows implies that $\sum_j P_{ij} \log P_{ij}$ is a constant c, independent of i, so $I(\mathcal{A},\mathcal{B}) = H(\mathcal{B}) + c$ since $\sum_i p_i = 1$. Now c is independent of $\mathcal{A}$, so maximising $I(\mathcal{A},\mathcal{B})$ is equivalent to maximising $H(\mathcal{B})$. By Theorem 3.10, $H(\mathcal{B})$ has maximum value $\log s$, so $C = I_{\max} = \log s + c$, attained when all q_j are equal; since $q_j = \sum_i p_i P_{ij}$, the condition on columns implies that this happens if all p_i are equal. For the r-ary symmetric channel, we obtain $C = \log s + c = \log r + P \log P + \overline{P} \log \overline{P} - \overline{P} \log(r-1)$. (When $r = 2$ this agrees with the value $1 - H(P)$ for the BSC.)

4.16 $I(\mathcal{A},\mathcal{B}) = H(\mathcal{B}) - H(\mathcal{B} \mid \mathcal{A}) = -q_1 \log q_1 - q_2 \log q_2 + p_1(P_{11} \log P_{11} + P_{12} \log P_{12}) + p_2(P_{21} \log P_{21} + P_{22} \log P_{22})$. The two linear equations $P_{i1}c_1 + P_{i2}c_2 = P_{i1} \log P_{i1} + P_{i2} \log P_{i2}$ for c_1, c_2 can be solved if $\det(P_{ij}) \neq 0$, or equivalently $P_{1j} \neq P_{2j}$ for $j = 1, 2$, and when this fails we can still solve them with $c_j = \log P_{1j} = \log P_{2j}$. Then $I = -q_1 \log q_1 - q_2 \log q_2 + p_1(P_{11}c_1 + P_{12}c_2) + p_2(P_{21}c_1 + P_{22}c_2)$, and since $p_1 P_{1j} + p_2 P_{2j} = q_j$ for $j = 1, 2$, we get $I = -q_1 \log q_1 - q_2 \log q_2 + q_1 c_1 + q_2 c_2$ as a function of q_1 and q_2. To maximise I subject to $q_1 + q_2 = 1$, define $\Phi = I + \lambda(q_1 + q_2 - 1)$ and solve $\partial\Phi/\partial q_1 = \partial\Phi/\partial q_2 = q_1 + q_2 - 1 = 0$. The first two equations give $c_j + \lambda = 1 + \log q_j$, so $c_1 - \log q_1 = c_2 - \log q_2$. Then $C = I_{\max} = q_1(c_1 - \log q_1) + q_2(c_2 - \log q_2) = c_j - \log q_j$ for $j = 1, 2$, using $q_1 + q_2 = 1$. Thus $2^C q_j = 2^{c_j}$, so $2^C = 2^C(q_1 + q_2) = 2^{c_1} + 2^{c_2}$ and hence $C = \log(2^{c_1} + 2^{c_2})$. If $P_{11} = P_{22}$ then Γ is the BSC, with $P_{11} = P_{22} = P$ and $P_{12} = P_{21} = \overline{P}$; the linear equations $Pc_1 + \overline{P}c_2 = -H(P) = \overline{P}c_1 + Pc_2$ give $c_1 = c_2 = -H(P)$ and $C = \log(2^{c_1} + 2^{c_2}) = 1 - H(P)$.

4.17 Let Γ_1, Γ_2 and $\Gamma = \Gamma_1 + \Gamma_2$ have r, r' and $r + r'$ input symbols and s, s' and $s + s'$ output symbols. Let $(p_1, \ldots, p_{r+r'})$ be an input distribution for Γ, with the symbols of Γ_1 ordered before those of Γ_2. If $u = p_1 + \cdots +$

p_r and $v = p_{r+1} + \cdots + p_{r+r'}$, so $u + v = 1$, then $(p_1/u, \ldots, p_r/u)$ and $(p_{r+1}/v, \ldots, p_{r+r'}/v)$ are input distributions for Γ_1 and Γ_2. If (p_i) gives output distribution (q_j) for Γ, then by linearity, $(p_1/u, \ldots, p_r/u)$ gives output distribution $(q_1/u, \ldots, q_s/u)$ for Γ_1; in particular, $q_1 + \cdots + q_s = u$. The output $\mathcal{B}_1$ of Γ_1 has entropy $H(\mathcal{B}_1) = -\sum_{i=1}^{r}(q_i/u)\log(q_i/u) = \log u - (1/u)\sum_{i=1}^{r} q_i \log q_i$, so $\sum_{i=1}^{r} q_i \log q_i = u \log u - uH(\mathcal{B}_1)$, with a similar result for Γ_2, giving $H(\mathcal{B}) = -u\log u - v\log v + uH(\mathcal{B}_1) + vH(\mathcal{B}_2)$ (the information $H(u)$ about which Γ_i is used, plus the weighted average of the output entropies of Γ_1 and Γ_2). Likewise $H(\mathcal{B} \mid \mathcal{A}) = uH(\mathcal{B}_1 \mid \mathcal{A}_1) + vH(\mathcal{B}_2 \mid \mathcal{A}_2)$, so $I(\mathcal{A}, \mathcal{B}) = -u\log u - v\log v + uI(\mathcal{A}_1, \mathcal{B}_1) + vI(\mathcal{A}_2, \mathcal{B}_2)$, with similar interpretations. We maximise $I(\mathcal{A}, \mathcal{B})$ by taking $I(\mathcal{A}_i, \mathcal{B}_i) = C_i$ (its maximum value) and then choosing u, v to maximise $I = -u\log u - v\log v + uC_1 + vC_2$ subject to $u + v = 1$. This is essentially the problem we faced in Exercise 4.16, so the method used there gives $C = \log(2^{C_1} + 2^{C_2})$. When $\Gamma_1 = \Gamma_2$ we get $C = C_1 + 1$, the extra unit of information indicated by which copy of Γ_1 is used.

4.18 $\Pr(a \mid c) = \sum_b \Pr(a \mid b)\Pr(b \mid c)$, so multiplying by $\Pr(c)$ and using $\Pr(c)\Pr(b \mid c) = \Pr(b, c)$ gives $\Pr(c)\Pr(a \mid c) = \sum_b \Pr(a \mid b)\Pr(b, c)$, and hence

$$\sum_b \sum_c \left(\Pr(b, c) \sum_a \Pr(a \mid b) \log \Pr(a \mid c) \right)$$

$$= \sum_c \left(\Pr(c) \sum_a \Pr(a \mid c) \log \Pr(a \mid c) \right)$$

$$= -H(\mathcal{A} \mid \mathcal{C}).$$

Also $\sum_c \Pr(b, c) = \Pr(b)$ implies that

$$\sum_b \sum_c \left(\Pr(b, c) \sum_a \Pr(a \mid b) \log \Pr(a \mid b) \right)$$

$$= \sum_b \left(\Pr(b) \sum_a \Pr(a \mid b) \log \Pr(a \mid b) \right)$$

$$= -H(\mathcal{A} \mid \mathcal{B}),$$

so

$$\sum_b \sum_c \left(\Pr(b, c) \sum_a \Pr(a \mid b) \big(\log \Pr(a \mid b) - \log \Pr(a \mid c) \big) \right)$$

$$= H(\mathcal{A} \mid \mathcal{C}) - H(\mathcal{A} \mid \mathcal{B}).$$

Corollary 3.9 shows that $\sum_a \Pr(a \mid b)\big(\log \Pr(a \mid b) - \log \Pr(a \mid c)\big) \geq 0$ for all b and c, so $H(\mathcal{A} \mid \mathcal{C}) \geq H(\mathcal{A} \mid \mathcal{B})$ and hence $I(\mathcal{A}, \mathcal{C}) =$

$H(\mathcal{A}) - H(\mathcal{A} \mid \mathcal{C}) \leq H(\mathcal{A}) - H(\mathcal{A} \mid \mathcal{B}) = I(\mathcal{A}, \mathcal{B})$. These inequalities show that further transmission (from $\mathcal{B}$ to $\mathcal{C}$) never decreases uncertainty about $\mathcal{A}$, and never increases mutual information about $\mathcal{A}$. We have $C = \max I(\mathcal{A}, \mathcal{C}) \leq \max I(\mathcal{A}, \mathcal{B}) = C_1$, and similarly $I(\mathcal{A}, \mathcal{C}) \leq I(\mathcal{B}, \mathcal{C})$ gives $C \leq C_2$, so $C \leq \min(C_1, C_2)$. If $\Gamma_1 = \Gamma_2$ is a BSC with capacity $C_1 = C_2 = 1 - H(P)$, then Exercise 4.12 shows that Γ is a BSC with probability $P' = (1 + (2P - 1)^2)/2$ and capacity $C = 1 - H(P')$. If $P = 0$ or 1 then $P' = 1$ and $C = C_1 = C_2 = 1$; if $P = \frac{1}{2}$ then $P' = \frac{1}{2}$ and $C = C_1 = C_2 = 0$. Otherwise, $|P' - \frac{1}{2}| < |P - \frac{1}{2}|$ giving $C < C_1 = C_2$.

Chapter 5

5.1 A decision rule is simply a function $\Delta : B \rightarrow A$, so there are $|A|^{|B|} = r^s$ decision rules.

5.2 We have

$$(R_{ij}) = \begin{pmatrix} pP & p\overline{P} \\ \overline{p}\overline{P} & \overline{p}P \end{pmatrix} = \begin{pmatrix} 0.72 & 0.18 \\ 0.02 & 0.08 \end{pmatrix},$$

and the greatest entry in each column is the first, so $\Delta(0) = \Delta(1) = 0$, giving $\mathrm{Pr}_E = 1 - \mathrm{Pr}_C = 1 - (0.72 + 0.18) = 0.1$.

5.3 For any decision rule $\Delta : B \rightarrow A$, $b_j \mapsto a_i = a_{j*}$,

$$\int_{\mathbf{p} \in \mathcal{P}} \mathrm{Pr}_C \, d\mathbf{p} = \int_{\mathbf{p} \in \mathcal{P}} \left(\sum_j p_{j*} P_{j*j} \right) d\mathbf{p} = \sum_j \left(P_{j*j} \int_{\mathbf{p} \in \mathcal{P}} p_{j*} \, d\mathbf{p} \right),$$

since each P_{j*j} is constant as $\mathbf{p}$ varies. Now $\int_{\mathbf{p} \in \mathcal{P}} p_{j*} \, d\mathbf{p}$ takes the same value for all j and Δ, since $\mathcal{P}$ is symmetric under all permutations of the coordinates p_i. Hence Δ maximises $\int_{\mathbf{p} \in \mathcal{P}} \mathrm{Pr}_C \, d\mathbf{p}$ if it maximises P_{j*j} for each j, and this is the maximum likelihood rule.

5.4 $d(\mathbf{u}, \mathbf{v}) = i$ if and only if $\mathbf{v}$ differs from $\mathbf{u}$ in exactly i coordinate positions; there are $\binom{n}{i}$ ways of choosing these positions, and for each coordinate position there are $r - 1$ different coordinates $\mathbf{v}$ can have, so there are $\binom{n}{i}(r-1)^i$ possibilities for $\mathbf{v}$. The Binomial Theorem gives $\sum_{i=0}^n \binom{n}{i}(r-1)^i = (r - 1 + 1)^n = r^n = |A^n|$.

5.5 The largest subsets with this property have four elements. They are the vertex-sets $\{000, 110, 101, 011\}$ and $\{100, 010, 001, 111\}$ of the two tetrahedra embedded in the cube $\mathbf{Z}_2^3$. In $\mathbf{Z}_2^n$ the largest such subsets have 2^{n-1} elements: there are two such sets, consisting of the words of length n with an even or an odd number of symbols 1.

5.6 Let $\mathbf{u}, \mathbf{v}, \mathbf{w} \in A^n$. If $u_i \neq w_i$ then $u_i \neq v_i$ or $v_i \neq w_i$, so $d(\mathbf{u}, \mathbf{w}) = |\{i \mid u_i \neq w_i\}| \leq |\{i \mid u_i \neq v_i \text{ or } v_i \neq w_i\}| \leq |\{i \mid u_i \neq v_i\}| + |\{i \mid v_i \neq w_i\}| = d(\mathbf{u}, \mathbf{v}) + d(\mathbf{v}, \mathbf{w})$.

5.7 Since $\lambda + \mu = 1$, the Binomial Theorem gives

$$1 = (\lambda + \mu)^n = \sum_{i=0}^{n} \binom{n}{i} \lambda^i \mu^{n-i} \geq \sum_{i \leq \lambda n} \binom{n}{i} \lambda^i \mu^{n-i} \geq \sum_{i \leq \lambda n} \binom{n}{i} \lambda^{\lambda n} \mu^{\mu n};$$

the last inequality is because $\lambda/\mu \leq 1$ and $i \leq \lambda n$ imply that

$$\lambda^i \mu^{n-i} = \left(\frac{\lambda}{\mu}\right)^i \mu^n \geq \left(\frac{\lambda}{\mu}\right)^{\lambda n} \mu^n = \lambda^{\lambda n} \mu^{n-\lambda n} = \lambda^{\lambda n} \mu^{\mu n}.$$

Dividing by $\lambda^{\lambda n} \mu^{\mu n}$ gives $\sum_{i \leq \lambda n} \binom{n}{i} \leq \lambda^{-\lambda n} \mu^{-\mu n} = (\lambda^{-\lambda} \mu^{-\mu})^n$, so

$$\log_2 \sum_{i \leq \lambda n} \binom{n}{i} \leq n(-\lambda \log_2 \lambda - \mu \log_2 \mu) = nH_2(\lambda)$$

and hence $\sum_{i \leq \lambda n} \binom{n}{i} \leq 2^{nH_2(\lambda)}$.

5.8 $\mathcal{R}_n = \{\mathbf{0} = 00\ldots0, \mathbf{1} = 11\ldots1\}$. The received word $\mathbf{v}$ consists of n symbols, equal to $0, ?$ or $1, ?$ as $\mathbf{u} = \mathbf{0}$ or $\mathbf{1}$ was transmitted, so let $\Delta(\mathbf{v}) = \mathbf{0}$ or $\mathbf{1}$ if $\mathbf{v}$ contains a letter 0 or 1, and let $\Delta(\mathbf{v})$ be undefined if $\mathbf{v} = ??\ldots?$. Then decoding is correct unless $\mathbf{v} = ??\ldots?$, so $\mathrm{Pr}_E = \mathrm{Pr}(\mathbf{v} = ??\ldots?) = \overline{P}^n \to 0$ as $n \to \infty$.

5.9 The channel matrix is $\left(\begin{smallmatrix} 1 & 0 \\ Q & P \end{smallmatrix}\right)$. Since $1 > Q$ and $P > 0$, the maximum likelihood rule is $\Delta(0) = 0$, $\Delta(1) = 1$, with $\mathrm{Pr}_C = p + \overline{p}P$, $\mathrm{Pr}_E = \overline{p}Q$. If 000 is transmitted, it is received correctly. If 111 is transmitted, it is received with $0, 1, 2$ or 3 errors, with probabilities $P^3, 3P^2Q, 3PQ^2$ and Q^3. Since $1 > Q^3$ and $P^3, 3P^2Q, 3PQ^2 > 0$, the maximum likelihood rule gives $\Delta(000) = 000$ and $\Delta(\mathbf{v}) = 111$ for all $\mathbf{v} \neq 000$. This differs from majority and nearest neighbour decoding, since (for example) $\Delta(100) = 111$ and not 000. The maximum likelihood rule gives $\mathrm{Pr}_C = p + \overline{p}(1 - Q^3)$, $\mathrm{Pr}_E = \overline{p}Q^3$, and the rate is $R = 1/3$. If $\mathcal{R}_n = \{00\ldots0, 11\ldots1\}$ is used, $\mathrm{Pr}_E = \overline{p}Q^n$ and $R = 1/n$; both approach 0 as $n \to \infty$.

5.10 If d denotes $d(\mathbf{u}, \mathbf{v})$, the forward probability is $\mathrm{Pr}(\mathbf{v}|\mathbf{u}) = P^{n-d}Q^d = P^n(Q/P)^d$; since $Q/P < 1$ this decreases as d increases, so the maximum likelihood rule (given $\mathbf{v}$, maximise $\mathrm{Pr}(\mathbf{v}|\mathbf{u})$) minimises d, and hence agrees with nearest neighbour decoding Δ. If w denotes $d(\mathbf{0}, \mathbf{v})$ then $d(\mathbf{1}, \mathbf{v}) = n - w$, so $\Delta(\mathbf{v}) = \mathbf{0}$ or $\mathbf{1}$ as $w < n - w$ or $w > n - w$; now

$\mathbf{v}$ has w symbols $v_i = 1$ and $n - w$ symbols $v_i = 0$, so Δ agrees with majority decoding. Using this rule Δ, and putting $n = 2t + 1$, we have

$$\mathrm{Pr_E} = \mathrm{Pr}\,(> t \text{ errors})$$

$$= \binom{2t+1}{t+1} P^t Q^{t+1} + \binom{2t+1}{t+2} P^{t-1} Q^{t+2} + \cdots + \binom{2t+1}{2t+1} P^0 Q^{2t+1}$$

$$\leq (t+1) \binom{2t+1}{t+1} P^t Q^{t+1}$$

$$= \frac{(2t+1)!}{(t!)^2} P^t Q^{t+1} \quad (= a_t, \text{ say})$$

since there are $t + 1$ summands, and the greatest is the first since $Q/P < 1$ and the binomial coefficients are decreasing. As $t \to \infty$,

$$\frac{a_{t+1}}{a_t} = \frac{(2t+3)(2t+2)}{(t+1)^2} PQ \to 4PQ < 1,$$

since $PQ = P - P^2 < \frac{1}{4}$ for $\frac{1}{2} < P \leq 1$. Thus $a_t \to 0$ as $t \to \infty$, so $\mathrm{Pr_E} \to 0$ as $n \to \infty$. The rate $R = \frac{1}{n} \to 0$ as $n \to \infty$, whereas Shannon's Theorem requires $R \approx C > 0$, so this does not prove the theorem.

5.11 Each toss multiplies the current capital by 2λ or 2μ as Γ transmits the outcome correctly or incorrectly, so after m correct and $n - m$ incorrect transmissions the initial capital is multiplied by $(2\lambda)^m (2\mu)^{n-m} = 2^n \lambda^m \mu^{n-m}$. Hence $c_n = 2^n \lambda^m \mu^{n-m} c_0$, and so $\frac{1}{n} \log(c_n/c_0) = 1 + \frac{m}{n} \log \lambda + \frac{n-m}{n} \log \mu$. By the Law of Large Numbers (Appendix B), we can expect $m/n \approx P$ and $(n - m)/n \approx Q$ with probability approaching 1 as $n \to \infty$, so $G \approx 1 + P \log \lambda + Q \log \mu$. Maximising G is equivalent to choosing λ, μ to minimise $-P \log \lambda - Q \log \mu$, and by Corollary 3.9 this is achieved by taking $\lambda = P$ (so that $\mu = Q$), with $G \approx 1 + P \log P + Q \log Q = 1 - H(P) = C$. If $\frac{1}{2} < P < 1$ then using a repetition code (as in §5.2) has the effect of reducing the error-probability of Γ, thus increasing C and G.

5.12 If b_j is received, the gambler bets a proportion λ_{ij} of his capital on each a_i, where $\sum_i \lambda_{ij} = 1$. If the input is a_i, this multiplies his capital by λ_{ij}/p_i, so after n bets $c_n = \prod_i \prod_j (\lambda_{ij}/p_i)^{m_{ij}} c_0$, where m_{ij} is the number of times a_i is transmitted and b_j is received. Thus $G = \lim_{n \to \infty} \frac{1}{n} \log(c_n/c_0) = \sum_i \sum_j \lim_{n \to \infty} (m_{ij}/n) \log(\lambda_{ij}/p_i)$. The Law of Large Numbers gives $m_{ij}/n \approx R_{ij}$ with probability approaching 1 as $n \to \infty$, so $G \approx \sum_i \sum_j R_{ij} \log(\lambda_{ij}/p_i) = \sum_i \sum_j R_{ij} \log \lambda_{ij} - \sum_i p_i \log p_i = \sum_j (\sum_i R_{ij} \log \lambda_{ij}) + H(\mathcal{A})$. Given $\mathcal{A}$ and Γ, the gambler can maximise G by maximising $\sum_i R_{ij} \log \lambda_{ij}$ for each j. Since $\sum_i R_{ij} = q_j$ for each j, Corollary 3.9 implies that this is achieved by taking $\lambda_{ij} = R_{ij}/q_j = Q_{ij}$, so $G \approx \sum_i \sum_j R_{ij} \log Q_{ij} + H(\mathcal{A}) = -H(\mathcal{A} \mid \mathcal{B}) + H(\mathcal{A}) = I(\mathcal{A}, \mathcal{B})$. The maximum value this can take (as (p_i) varies) is the capacity C of Γ. If a successful bet regains $1/p_i'$ times the stake, we replace λ_{ij}/p_i with λ_{ij}/p_i' above, so $\lambda_{ij} = Q_{ij}$ again, giving an exponential growth rate $G' \approx -H(\mathcal{A} \mid \mathcal{B}) - \sum_i p_i \log p_i' \geq I(\mathcal{A}, \mathcal{B})$ by Corollary 3.9; thus the gambler is generally better off (equivalently, the bookmakers choose the odds $1/p_i$ to minimise their losses).

Chapter 6

6.1 $\mathcal{C} \cap \mathcal{C}'$ and $\mathcal{C} + \mathcal{C}'$ are non-empty and closed under linear combinations, so they are linear. If $\mathcal{C} \subseteq \mathcal{C}'$ or $\mathcal{C}' \subseteq \mathcal{C}$ then $\mathcal{C} \cup \mathcal{C}'$ is $\mathcal{C}'$ or $\mathcal{C}$ and hence is linear; if $\mathcal{C} \nsubseteq \mathcal{C}'$ and $\mathcal{C}' \nsubseteq \mathcal{C}$ then $\mathcal{C} \cup \mathcal{C}'$ is not linear, for if $c \in \mathcal{C} \setminus \mathcal{C}'$ and $c' \in \mathcal{C}' \setminus \mathcal{C}$ then $c, c' \in \mathcal{C} \cup \mathcal{C}'$ but $c + c' \notin \mathcal{C} \cup \mathcal{C}'$.

6.2 If $\mathbf{a} = 1101$ then $\mathbf{u} = 1010101$. If $\mathbf{v} = 1010111$ is received then $\mathbf{s} = 110$, representing 6 and indicating an incorrect 6th symbol, so $\Delta(\mathbf{v}) = 1010101 = \mathbf{u}$. If $\mathbf{v}' = 1011111$ is received then $\mathbf{s}' = 010$, representing 2, so $\Delta(\mathbf{v}') = 1111111 \neq \mathbf{u}$.

6.3 Taking all linear combinations of the basis vectors $\mathbf{u}_i$ in Example 6.5, we get

$$\mathcal{H}_7 = \{0000000, 1110000, 1001100, 0101010, 1101001, 0111100,$$
$$1011010, 0011001, 1100110, 0100101, 1000011, 0010110,$$
$$1010101, 0110011, 0001111, 1111111\}.$$

By inspection, the minimum weight of a non-zero code-word is 3, so $d = 3$.

6.4 The elements of $\overline{\mathcal{C}}$ are $\overline{\mathbf{u}} = u_1 \ldots u_{n+1}$, where $\mathbf{u} = u_1 \ldots u_n \in \mathcal{C}$ and $u_{n+1} = u_1 + \cdots + u_n$ in $\mathbf{Z}_2$. Thus $\mathrm{wt}(\overline{\mathbf{u}}) = \mathrm{wt}(\mathbf{u})$ or $\mathrm{wt}(\mathbf{u}) + 1$ as $\mathrm{wt}(\mathbf{u})$ is even or odd, so by Lemma 6.8 $\overline{\mathcal{C}}$ has minimum distance d or $d + 1$ respectively. Taking $\mathcal{C} = \mathcal{H}_7$, with $d = 3$, Exercise 6.3 shows that $\overline{\mathcal{H}_7}$ has code-words

$$00000000, 11100001, 10011001, 01010101, 11010010, 01111000,$$
$$10110100, 00110011, 11001100, 01001011, 10000111, 00101101,$$
$$10101010, 01100110, 00011110, 1111111,$$

so it has minimum distance 4.

6.5 Both properties are equivalent to the condition that, for some t, every word is at distance at most t from a unique code-word.

6.6 $\mathcal{H}_7$ is perfect, with $t = 1$, so decoding is correct if and only if there is at most one error. This has probability $P^7 + 7P^6Q = -6P^7 + 7P^6$, so $\mathrm{Pr}_E = 1 + 6P^7 - 7P^6$. For small Q, the Binomial Theorem gives $P^i = (1 - Q)^i \approx 1 - iQ + \binom{i}{2}Q^2$, so $\mathrm{Pr}_E \approx 21Q^2$.

6.7 $\overline{\mathcal{H}_7}$ has $d = 4$ by Exercise 6.4, so $t = 1$ and $\sum_{i=0}^{t} \binom{n}{i}(q-1)^i = 1 + \binom{8}{1} = 9$. Thus the $2^4 = 16$ spheres $S_t(\mathbf{u})$ cover only $16 \times 9 = 144$ of the $2^8 = 256$ vectors $\mathbf{v} \in V = F_2^8$, so $\overline{\mathcal{H}_7}$ is not perfect.

6.8 Apply Stirling's approximation $m! \sim (m/e)^m \sqrt{2\pi m}$ (see [Fi83] or [La83]) to the three factorials in $\binom{n}{t} = n!/t!(n-t)!$, and then take logarithms.

6.9 If $d = 3$ then $t = \lfloor \frac{d-1}{2} \rfloor = 1$ by Theorem 6.15, so putting $q = 3$ in Theorem 6.15 gives $A_3(n,3) \le \lfloor 3^n/(2n+1) \rfloor$. If $n = 3,4,5,6,7,\ldots$ then $A_3(n,3) \le 3, 9, 22, 56, 145, \ldots$. If $d = 4$ then $t = 1$, so Theorem 6.15 gives $A_2(n,4) \le \lfloor 2^n/(n+1) \rfloor$ as in Example 6.16. If $d = 5$ then $t = 2$, so $A_2(n,5) \le \lfloor 2^n/(1 + n + \binom{n}{2})) \rfloor = \lfloor 2^{n+1}/(n^2 + n + 2) \rfloor$.

6.10 Example 6.22 gives $A_2(4,3) = 2$ or 3. If $\mathcal{C} = \{\mathbf{u}, \mathbf{v}, \mathbf{w}\}$ is a binary code with $n = 4$ and $d = 3$, then $\mathbf{v}$ and $\mathbf{w}$ each differ from $\mathbf{u}$ in at least three of their four coordinate positions; at least two of these coordinate positions i and j must be the same, so $v_i \ne u_i \ne w_i$ and $v_j \ne u_j \ne w_j$; since the code is binary this forces $v_i = w_i$ and $v_j = w_j$, so $d(\mathbf{v}, \mathbf{w}) \le 2$, contradicting $d = 3$. Thus $A_2(4,3) < 3$, so $A_2(4,3) = 2$. The code $\{0000, 1110\}$ attains this bound.

6.11 Theorem 6.15 with $q = d = 3$ gives $A_3(n,3) \ge \lceil 3^n/(1 + 2\binom{n}{1} + 2^2\binom{n}{2})) \rceil = \lceil 3^n/(2n^2 + 1) \rceil$.

6.12 For $n = 1$, $H = (1)$ or $(-)$. For $n = 2$, $H = \pm\left(\begin{smallmatrix} 1 & 1 \\ 1 & - \end{smallmatrix}\right), \pm\left(\begin{smallmatrix} 1 & 1 \\ - & 1 \end{smallmatrix}\right), \pm\left(\begin{smallmatrix} 1 & - \\ 1 & 1 \end{smallmatrix}\right)$ or $\pm\left(\begin{smallmatrix} 1 & - \\ - & - \end{smallmatrix}\right)$.

6.13 The entries of H' are all ± 1, since those of H are, and it is easy to check that distinct rows of H' are orthogonal.

6.14 $1111, 0000, 1010, 0101, 1100, 0011, 1001, 0110$. These form the binary parity-check code $\mathcal{P}_4$, which is linear (Example 6.4).

6.15 Applying Lemma 6.24 to the Hadamard matrix H of order 4 in Example 6.26, we get a Hadamard matrix of order 8

$$H' = \begin{pmatrix} 1 & 1 & 1 & 1 & 1 & 1 & 1 & 1 \\ 1 & - & 1 & - & 1 & - & 1 & - \\ 1 & 1 & - & - & 1 & 1 & - & - \\ 1 & - & - & 1 & 1 & - & - & 1 \\ 1 & 1 & 1 & 1 & - & - & - & - \\ 1 & - & 1 & - & - & 1 & - & 1 \\ 1 & 1 & - & - & - & - & 1 & 1 \\ 1 & - & - & 1 & - & 1 & 1 & - \end{pmatrix}$$

giving 16 code-words 11111111, 00000000, 10101010, 01010101, 11001100, 00110011, 10011001, 01100110, 11110000, 00001111, 10100101, 01011010, 11000011, 00111100, 10010110 and 01101001. The rate is $(\log_2 16)/8 = 1/2$. Since $d = 4$, Theorem 6.10 gives $t = 1$; the code detects $d - 1 = 3$ errors.

6.16 A cubic $f(x)$ is irreducible if and only if it has no linear factors, i.e. no roots, so $f(x) = x^3 + x + 1$ and $g(x) = x^3 + x^2 + 1$ are the only possibilities. If α and β are roots of f and g, then $F = \{a\alpha^2 + b\alpha + c \mid a, b, c \in \mathbf{Z}_2\}$ and $F' = \{a\beta^2 + b\beta + c \mid a, b, c \in \mathbf{Z}_2\}$ are fields of order 8, with $\alpha^3 = \alpha + 1$ and $\beta^3 = \beta^2 + 1$. Then $(\beta + 1)^3 = (\beta + 1) + 1$, so $a\alpha^2 + b\alpha + c \mapsto a(\beta + 1)^2 + b(\beta + 1) + c = a\beta^2 + b\beta + (a + b + c)$ is an isomorphism $F \to F'$.

6.17 If $f(x) = x^2 + 1$ has a root $\alpha \in \mathbf{Z}_p$, then $\alpha^2 = -1 \neq 1$ but $\alpha^4 = (-1)^2 = 1$, so α has order 4 in the multiplicative group $\mathbf{Z}_p^* = \mathbf{Z}_p \setminus \{0\}$; thus 4 divides $|\mathbf{Z}_p^*| = p - 1$, impossible since $p \equiv 3 \bmod (4)$. Hence a root α of f is not in $\mathbf{Z}_p$, so $F = \{a\alpha + b \mid a, b \in \mathbf{Z}_p\}$ is a field of order p^2, with $\alpha^2 = -1$. A similar argument, using $x^3 - 1 = (x - 1)(x^2 + x + 1)$, shows that $x^2 + x + 1$ is irreducible over $\mathbf{Z}_p$ for $p \equiv 2 \bmod (3)$, in which case there is a field $F = \{a\alpha + b \mid a, b \in \mathbf{Z}_p\}$ of order p^2 with $\alpha^2 + \alpha + 1 = 0$.

6.18 Since all code-words differ in at least d positions, deleting $d - 1$ symbols gives a set of M distinct words of length $n - d + 1$ over F_q. There are at most q^{n-d+1} such words, so $M \leq q^{n-d+1}$. Now take logarithms. A repetition code $\mathbf{R}_n$ attains this bound, with $M = q$ and $d = n$, as does a parity-check code $\mathcal{P}_n$, with $M = q^{n-1}$ and $d = 2$.

6.19 $\binom{23}{0} + \binom{23}{1} + \binom{23}{2} + \binom{23}{3} = 2048 = 2^{11}$, so $2^{12}(\binom{23}{0} + \binom{23}{1} + \binom{23}{2} + \binom{23}{3}) = 2^{23}$; this gives equality in Hamming's sphere-packing bound with $n = 23$, $q = 2$, $t = 3$, $M = 2^{12}$. Similarly $\binom{11}{0} + \binom{11}{1}.2 + \binom{11}{2}.2^2 = 243 = 3^5$ gives equality with $n = 11$, $q = 3$, $t = 2$, $M = 3^6$. This suggests the existence of 12- and 6-dimensional perfect linear codes with these parameters; these *Golay codes* are described in §7.5. (However, see Exercise 7.16.)

6.20 $\mathcal{C}_1 \oplus \mathcal{C}_2$ and $\mathcal{C}_1 * \mathcal{C}_2$ are subsets of the $2n$-dimensional vector space $\mathcal{V} \oplus \mathcal{V}$, so they are codes of length $2n$. The M_1, M_2 vectors $\mathbf{x}, \mathbf{y}$ give rise to $M_1 M_2$ distinct vectors $(\mathbf{x}, \mathbf{y})$ or $(\mathbf{x}, \mathbf{x} + \mathbf{y})$ respectively, so each code contains $M_1 M_2$ code-words. Elements $(\mathbf{x}, \mathbf{y})$ and $(\mathbf{x}', \mathbf{y}')$ of $\mathcal{C}_1 \oplus \mathcal{C}_2$ are distinct if and only if $\mathbf{x} \neq \mathbf{x}'$ or $\mathbf{y} \neq \mathbf{y}'$, in which case $d((\mathbf{x}, \mathbf{y}), (\mathbf{x}', \mathbf{y}')) = d(\mathbf{x}, \mathbf{x}') + d(\mathbf{y}, \mathbf{y}') \geq \min(d_1, d_2)$; this bound is attained by taking $\mathbf{x} = \mathbf{x}'$ or $\mathbf{y} = \mathbf{y}'$ and letting the other (distinct) pair be as close as possible, so $d(\mathcal{C}_1 \oplus \mathcal{C}_2) = \min(d_1, d_2)$. In $\mathcal{C}_1 * \mathcal{C}_2$, if $\mathbf{x} = \mathbf{x}'$ and $\mathbf{y} \neq \mathbf{y}'$ then $d((\mathbf{x}, \mathbf{x}+\mathbf{y}), (\mathbf{x}', \mathbf{x}'+\mathbf{y}')) = d(\mathbf{y}, \mathbf{y}')$ has minimum value d_2, and if $\mathbf{x} \neq \mathbf{x}'$ and $\mathbf{y} = \mathbf{y}'$ then $d((\mathbf{x}, \mathbf{x} + \mathbf{y}), (\mathbf{x}', \mathbf{x}' + \mathbf{y}')) = 2d(\mathbf{x}, \mathbf{x}')$ has minimum value $2d_1$; if $\mathbf{x} \neq \mathbf{x}'$ and $\mathbf{y} \neq \mathbf{y}'$ then $d(\mathbf{x}, \mathbf{x}') \geq d_1$ and $d(\mathbf{x}+\mathbf{y}, \mathbf{x}'+\mathbf{y}') \geq |d_1 - d_2|$, so $d((\mathbf{x}, \mathbf{x}+\mathbf{y}), (\mathbf{x}', \mathbf{x}'+\mathbf{y}')) \geq d_1 + |d_1 - d_2| \geq d_2 \geq \min(2d_1, d_2)$, and thus $d(\mathcal{C}_1 * \mathcal{C}_2) = \min(2d_1, d_2)$. If each $\mathcal{C}_i$ is linear, then $\mathcal{C}_1 \oplus \mathcal{C}_2$ and $\mathcal{C}_1 * \mathcal{C}_2$ are linear subspaces of $\mathcal{V} \oplus \mathcal{V}$ (closure is easily checked), so they are linear codes; they have dimension $\log_q M_1 M_2 = \log_q M_1 + \log_q M_2 = k_1 + k_2$.

6.21 If the j-th digit a_j is changed to $b_j \neq a_j$, then $\sum_{i=1}^{10} i a_i \, (\equiv 0 \bmod (11))$ is replaced with $\sum_{i \neq j} i a_i + j b_j = \sum_i i a_i + j(b_j - a_j) \equiv 0 + j(b_j - a_j) \equiv j(b_j - a_j)$, and this $\not\equiv 0$ since $j, b_j - a_j \not\equiv 0$, so the error is detected. Similarly, if a_j and a_k are transposed, where $a_j \neq a_k$, then $\sum_i i a_i$ is replaced with $\sum_i i a_i + j(a_k - a_j) + k(a_j - a_k) \equiv 0 + (j - k)(a_k - a_j) \not\equiv 0$, so the error is detected. In each case, it is important that 11 is prime, so that $x, y \not\equiv 0$ implies $xy \not\equiv 0$ (false for composite moduli). 3-540-76197-7 has $a_1 + 2a_2 + \cdots + 10a_{10} = 308 \equiv 0 \bmod (11)$, so it is a valid ISBN (in fact, of the SUMS textbook *Elementary Number Theory*, by Jones and Jones); the second and third differ from this by a transposition and a single error, so they cannot be ISBNs.

Chapter 7

7.1 Form $\overline{G}$ by adding an extra column to G so that each row-sum is 0. Form $\overline{H}$ from H by adding a column of $c = n - k$ entries 0, and then a row of $n + 1$ entries 1.

7.2 Form a generator matrix G for $\mathcal{C}_1 + \mathcal{C}_2$ by adjoining the rows of G_2 to G_1 and then using elementary row operations to eliminate linearly dependent rows. A similar process with the rows of H_1 and H_2 gives a parity-check matrix H for $\mathcal{C}_1 \cap \mathcal{C}_2$.

7.3 Each row of G_1 is a linear combination of rows of G_2, so $C_1 \subseteq C_2$; since $\dim C_1 = \dim C_2$, $C_1 = C_2$. Alternatively $G_2 H^T = O$, where H is the parity-check matrix for $C_1 = \mathcal{H}_7$ in Example 7.13, so $C_2 \subseteq C_1$; comparing dimensions gives equality.

7.4 Since $\mathcal{H}_n$ is a 1-error-correcting perfect code, nearest neighbour decoding corrects all error patterns with at most one error, but no others; the probability of no errors is P^n and the probability of a single error in a given position is $P^{n-1}Q$, so $\Pr_C = P^n + nP^{n-1}Q$ and $\Pr_E = 1 - \Pr_C = 1 - P^n - nP^{n-1}Q$. If $P < 1$ then P^n, $nP^{n-1} \to 0$ as $n \to \infty$, so $\Pr_E \to 1$. If $P = 1$ then $\Pr_E = 0$ for all n.

7.5 $\mathbf{u}H^T = \mathbf{0}$, so $\mathbf{u} \in \mathcal{H}_7$. The syndrome of $\mathbf{v}$ is $\mathbf{s} = \mathbf{v}H^T = 101 = \mathbf{c}_2^T$, indicating an error in position 2, so $\Delta(\mathbf{v}) = \mathbf{v} - \mathbf{e}_2 = 1100110$; this is $\mathbf{u}$, so decoding is correct. However $\mathbf{v}'$ has syndrome $\mathbf{s}' = \mathbf{v}'H^T = 110 = \mathbf{c}_3^T$, indicating an error in position 3; this gives $\Delta(\mathbf{v}') = \mathbf{v}' - \mathbf{e}_3 = 0010110 \neq \mathbf{u}$, so decoding is incorrect. This is because $\mathbf{v}'$ involves two errors, whereas $\mathbf{v}$ involves only one, and $\mathcal{H}_7$ corrects one error but not two.

7.6 The syndrome $\mathbf{s} = \mathbf{v}H^T = 010$ is the binary representation of 2, indicating an error in position 2; thus $\Delta(\mathbf{v}) = \mathbf{v} - \mathbf{e}_2 = 0111100 = \mathbf{u}$, so decoding is correct.

7.7 No vector can be a multiple of another, so they must generate distinct 1-dimensional subspaces. The number of such subspaces in $\mathcal{W}$ is $n = (q^c - 1)/(q-1)$, so this is the maximum number of vectors. If they are the columns of H, the corresponding code C has length n and dimension $k = n - c$. No two columns are linearly dependent, but three are (the sum of any two is a multiple of a third), so $d = 3$ by Theorem 7.27, giving $t = 1$ by Theorem 6.10. Then $\sum_{i=0}^{t} \binom{n}{i}(q-1)^i = 1 + n(q-1) = q^c = q^{n-k}$, so C is perfect.

7.8 Each $s \in S$ lies in $\binom{23}{4}$ 5-element subsets of S, each of which is contained in a unique block; conversely, each block containing s has $\binom{7}{4}$ 5-element subsets containing s, so s lies in $\binom{23}{4}/\binom{7}{4} = 253$ blocks. Similarly, the number of blocks containing each pair, triple and quadruple is $\binom{22}{3}/\binom{6}{3} = 77$, $\binom{21}{2}/\binom{5}{2} = 21$, and $\binom{20}{1}/\binom{4}{1} = 5$.

7.9 The given generator matrix is
$$\begin{pmatrix} 0 & 1 & 1 & 0 & 1 & 1 \\ 1 & 0 & 1 & 1 & 0 & 1 \\ 1 & 1 & 1 & 0 & 0 & 0 \end{pmatrix}.$$

Adding row 3 to rows 1 and 2, and then adding the new rows 1 and 2 to row 3, we get

$$G = \begin{pmatrix} 1 & 0 & 0 & 0 & 1 & 1 \\ 0 & 1 & 0 & 1 & 0 & 1 \\ 0 & 0 & 1 & 1 & 1 & 0 \end{pmatrix}, \quad \text{so} \quad H = \begin{pmatrix} 0 & 1 & 1 & 1 & 0 & 0 \\ 1 & 0 & 1 & 0 & 1 & 0 \\ 1 & 1 & 0 & 0 & 0 & 1 \end{pmatrix}.$$

110 is encoded as $\mathbf{c} = 110.G = 110110$, with $\mathbf{c}.H^{\mathrm{T}} = 000 = \mathbf{0}$. Since $n = 6$ and $k = 3$, the rate is $R = k/n = 1/2$. The minimum distance d is the minimum number of linearly dependent columns of H, and this is 3 (columns 1, 2, 3). The syndrome table

$\mathbf{v}_i =$	000000	100000	010000	001000	000100	000010	000001	100100
$\mathbf{s}_i =$	000	011	101	110	100	010	001	111

corrects all single-error patterns, and one double-error pattern 100100, so if each symbol has probability P, Q of correct/incorrect transmission then $\mathrm{Pr_E} = 1 - (P^6 + 6P^5Q + P^4Q^2)$.

7.10 Each $\mathcal{C}$ has an ordered basis $\mathbf{u}_1, \ldots, \mathbf{u}_k$. There are $q^n - 1$ choices for $\mathbf{u}_1 \in V$ (excluding $\mathbf{0}$), then $q^n - q$ for $\mathbf{u}_2$ (excluding multiples of $\mathbf{u}_1$), $\ldots, q^n - q^{k-1}$ choices for $\mathbf{u}_k$, hence $(q^n - 1) \ldots (q^n - q^{k-1})$ such bases in V. Similarly each $\mathcal{C}$ has $(q^k - 1) \ldots (q^k - q^{k-1})$ ordered bases, so the number of codes $\mathcal{C}$ is $(q^n - 1) \ldots (q^n - q^{k-1})/(q^k - 1) \ldots (q^k - q^{k-1})$.

7.11 $L_1 \cap L_2$ is a single point p, which lies on a unique third line L_3; the three sets $L_i \setminus \{p\}$ partition $S \setminus \{p\}$, so $L_1 + L_2 = (L_1 \setminus \{p\}) \cup (L_2 \setminus \{p\}) = (S \setminus \{p\}) \setminus (L_3 \setminus \{p\}) = S \setminus L_3$, the complement $\overline{L}_3$ of L_3. Thus the subspace $\mathcal{C}$ spanned by the lines contains the seven lines L and their seven complements $\overline{L}$, together with $L + L = \emptyset$ and $L + \overline{L} = S$. This set of sixteen subsets of S is closed under addition (for instance $L_1 + \overline{L}_2 = L_3$ and $\overline{L}_1 + \overline{L}_2 = \overline{L}_3$), so it is the whole of $\mathcal{C}$. Thus $\mathcal{C}$ is a binary linear code of length $n = 7$ and dimension $k = \log_2 16 = 4$. The non-zero code-words $L, \overline{L}$ and S have weight $|L| = 3, |\overline{L}| = 4$ and $|S| = 7$, so $\mathcal{C}$ has minimum distance $d = 3$ by Lemma 6.8, and hence $t = 1$ by Theorem 6.10. In §7.4 we showed that any two binary linear 1-error-correcting [7, 4]-codes are equivalent, so $\mathcal{C}$ is equivalent to $\mathcal{H}_7$.

7.12 Any pair of points form the support of a vector $\mathbf{v}$ of weight 2; since $\mathcal{C}$ is perfect with $t = 1$, $\mathbf{v}$ is at distance 1 from a unique code-word $\mathbf{u}$ of weight 3, whose support is the unique block containing the pair. Thus the code-words of weight 3 are the blocks of a Steiner system $S(2, 3, n)$. In $\mathcal{H}_n$, the coordinate positions $i = 1, \ldots, n = 2^c - 1$, written in binary notation to form the columns of the parity-check matrix H, consist of the non-zero vectors in F_2^c, so they correspond to the points of $PG(c-1,2)$; code-words of weight 3 correspond to the relations $\mathbf{c}_i + \mathbf{c}_j + \mathbf{c}_k = \mathbf{0}$ between columns of H (see §7.3), and hence to the lines $\{\mathbf{c}_i, \mathbf{c}_j, \mathbf{c}_k\}$ of $PG(c-1,2)$.

7.13 The identity permutation maps C to itself, and if permutations g and h do then so do gh and g^{-1}; thus $\text{Aut}(C)$ is a subgroup of S_n. Both $\mathcal{R}_n$ and $\mathcal{P}_n$ are invariant under all permutations, so they have automorphism group S_n. The code $\mathcal{R}_2 \oplus \mathcal{R}_2 = \{0000, 1100, 0011, 1111\}$ has eight automorphisms (12), (34), $(12)(34)$, $(13)(24)$, $(14)(23)$, (1324), (1423) and the identity (forming a dihedral group). The codes equivalent to C are those formed by applying a permutation to the n coordinates; two permutations yield the same code if and only if they lie in the same coset of $\text{Aut}(C)$ in S_n, so the number of equivalent codes is the number of cosets, namely $|S_n|/|\text{Aut}(C)| = n!/|\text{Aut}(C)|$. If $C = \mathcal{R}_2 \oplus \mathcal{R}_2$ there are $4!/8 = 3$ equivalent codes, namely C, $\{0000, 1010, 0101, 1111\}$ and $\{0000, 0110, 1001, 1111\}$.

7.14 By Exercises 7.11 and 7.12, any automorphism of $PG(2,2)$ induces an automorphism of $\mathcal{H}_7$, and vice versa, so their automorphism groups are isomorphic. The automorphisms of $PG(2,2)$ are induced by those of the corresponding vector space F_2^3, and these form the general linear group $GL(3,2)$ of invertible 3×3 matrices over F_2; only the identity matrix induces the identity automorphism of $PG(2,2)$, so $\text{Aut}(\mathcal{H}_7) \cong \text{Aut}(PG(2,2)) \cong GL(3,2)$. There are $2^3 - 1 = 7$ possibilities for the first row of a matrix in $GL(3,2)$; once this is chosen, there are $2^3 - 2 = 6$ possibilities for the second row, and then $2^3 - 2^2 = 4$ for the third, so $|\text{Aut}(\mathcal{H}_7)| = |GL(3,2)| = 7.6.4 = 168$. By Exercise 7.13 there are $7!/168 = 30$ codes equivalent to $\mathcal{H}_7$. Similarly if $n = 2^c - 1$ then $\text{Aut}(\mathcal{H}_n) \cong \text{Aut}(PG(c-1,2)) \cong GL(c,2)$, of order $(2^c-1)(2^c-2)(2^c-2^2) \ldots (2^c-2^{c-1})$, giving $n!/(2^c - 1)(2^c - 2)(2^c - 2^2) \ldots (2^c - 2^{c-1})$ equivalent codes.

7.15 Any $t+1$ points support a vector of weight $t+1$; since C is perfect, this is at distance t from a unique code-word of weight $d = 2t+1$, whose support is the unique block containing the $t+1$ points. Thus we have a Steiner system $S(t+1, d, n)$ (see Exercise 7.12 for the case $t = 1$). The number of blocks is $\binom{n}{t+1}/\binom{d}{t+1}$ (see §7.5), so $\binom{d}{t+1}$ divides $\binom{n}{t+1}$. Deleting i points for $i = 1, \ldots, t$ we obtain Steiner systems $S(t+1-i, d-i, n-i)$, so by the same argument $\binom{d-i}{t+1-i}$ divides $\binom{n-i}{t+1-i}$.

7.16 $1+90+\binom{90}{2} = 4096 = 2^{12}$, so the parameters $q = 2$, $n = 90$, $t = 2$, $M = 2^{78}$ give equality in Hamming's sphere-packing bound, suggesting the possible existence of a perfect 2-error-correcting binary code of length 90. However, if this exists then putting $d = 2t + 1 = 5$ and taking $i = 2$ in Exercise 7.15

we see that 3 divides 88, which is false. (Taking $i = 1$ also gives a slightly less obvious contradiction.)

7.17 Each coordinate position i contributes 1 or 0 to each side of the equation as just one of u_i, v_i is 1 or otherwise. The twelve rows of G are of length 24 and are independent, so they generate a binary linear $[24, 12]$-code $\mathcal{C}$. Each vertex of the icosahedron is adjacent to five others, so each row $\mathbf{r}$ of G has an even number $(1 + (12 - 5) = 8)$ of 1s, giving $\mathbf{r}.\mathbf{r} = 0$; similarly, any two distinct vertices have an even number of common non-neighbours, so the rows of G are mutually orthogonal and hence $\mathcal{C} \subseteq \mathcal{C}^\perp$; since $\dim(\mathcal{C}) = \dim(\mathcal{C}^\perp)$ we have $\mathcal{C} = \mathcal{C}^\perp$. Since P is binary and symmetric, $(P \mid I) = (-P^T \mid I)$; this is a parity-check matrix for $\mathcal{C}$, and hence a generator matrix for $\mathcal{C}^\perp = \mathcal{C}$. Each row of G has weight divisible by 4, and by the first result this property is preserved when elements $\mathbf{u}, \mathbf{v} \in \mathcal{C}$ are added, since self-duality implies that $c(\mathbf{u}, \mathbf{v})$ is always even. If $\mathbf{u} \in \mathcal{C}$ has x and y 1s in its first and last 12 entries, so that $x + y = \text{wt}(\mathbf{u})$, then $\mathbf{u}$ is a sum of x rows of G and also of y rows of G'; if $\text{wt}(\mathbf{u}) = 4$ then a sum of at most two rows of G (or equivalently of G') has weight 4, which is false by inspection. Thus $\mathcal{C}$ has minimum distance 8, and so the binary linear $[23, 12]$-code $\mathcal{C}^\circ$ has $d = 7$ and hence $t = 3$. Since $\sum_{i=0}^{3} \binom{23}{i} = 2^{23-12}$, $\mathcal{C}^\circ$ is perfect.

7.18 The six rows of G are independent and of length 12, so they generate a ternary linear $[12, 6]$-code. By inspection, the rows are mutually orthogonal, so $\mathcal{C} \subseteq \mathcal{C}^\perp$; comparing dimensions, we have $\mathcal{C} = \mathcal{C}^\perp$. Each $\mathbf{u} \in \mathcal{C}$ satisfies $\sum u_i^2 = 0$ in F_3 and hence has weight divisible by 3; it is therefore sufficient to show that $\text{wt}(\mathbf{u}) \neq 3$, and this follows by considering the various linear combinations of rows of G, so $\mathcal{C}$ has minimum distance 6. Then $\mathcal{C}^\circ$ is a ternary linear $[11, 6]$-code of minimum distance 5; it corrects 2 errors, and since $\sum_{i=0}^{2} \binom{11}{i} \cdot 2^i = 3^{11-6}$ it is perfect.

7.19 The basic properties of $\mathcal{RM}(r, m)$ follow from Exercise 6.20, using the inductive definition of this code. For instance, $\mathcal{RM}(0, m)$ and $\mathcal{RM}(m, m)$ are binary and linear, properties preserved by $*$, so every Reed–Muller code is binary and linear. Since $*$ doubles lengths, $\mathcal{RM}(r, m)$ has length $n = 2^m$. If $\mathcal{RM}(r, m - 1)$ and $\mathcal{RM}(r - 1, m - 1)$ have minimum distances $d_1 = 2^{m-1-r}$ and $d_2 = 2^{m-r}$ then $\mathcal{RM}(r, m)$ has minimum distance $d = \min(2d_1, d_2) = 2^{m-r}$. If $\mathcal{RM}(r, m - 1)$ and $\mathcal{RM}(r - 1, m - 1)$ have dimensions $k_1 = \sum_{i=0}^{r} \binom{m-1}{i}$ and $k_2 = \sum_{i=0}^{r-1} \binom{m-1}{i}$, then $\mathcal{RM}(r, m)$ has dimension

$$k_1 + k_2 = \binom{m-1}{0} + \sum_{i=1}^{r} \binom{m-1}{i} + \sum_{i=0}^{r-1} \binom{m-1}{i}$$

$$= \binom{m-1}{0} + \sum_{i=1}^{r} \binom{m-1}{i} + \sum_{i=1}^{r} \binom{m-1}{i-1}$$

$$= \binom{m}{0} + \sum_{i=1}^{r} \binom{m}{i} = \sum_{i=0}^{r} \binom{m}{i}.$$

$$\mathcal{RM}(1,2) = \mathcal{RM}(1,1) * \mathcal{RM}(0,1) = \{00,01,10,11\} * \{00,11\}$$

$$= \{0000,0011,0101,0110,1010,1001,1111,1100\}.$$

$$\mathcal{RM}(1,3) = \mathcal{RM}(1,2) * \mathcal{RM}(0,2) = \mathcal{RM}(1,2) * \{0000,1111\}$$

$$= \{00000000, 00110011, 01010101, 01100110,$$

$$10101010, 10011001, 11111111, 11001100,$$

$$00001111, 00111100, 01011010, 01101001,$$

$$10100101, 10010110, 11110000, 11000011\}.$$

Since $\dim \mathcal{RM}(1,3) = 4$, a basis consists of four independent code-words, such as $10010110, 01010101, 00110011, 00001111$, giving a generator matrix

$$G = \begin{pmatrix} 1 & 0 & 0 & 1 & 0 & 1 & 1 & 0 \\ 0 & 1 & 0 & 1 & 0 & 1 & 0 & 1 \\ 0 & 0 & 1 & 1 & 0 & 0 & 1 & 1 \\ 0 & 0 & 0 & 0 & 1 & 1 & 1 & 1 \end{pmatrix}.$$

This is not in systematic form, but interchanging columns 4 and 5 gives a generator matrix $G' = (I_4 \mid P)$ in systematic form, and hence a parity-check matrix $H' = (-P^T \mid I_4)$, for an equivalent code. Interchanging columns 4 and 5 of H' gives a parity-check matrix

$$H = \begin{pmatrix} 1 & 1 & 1 & 1 & 0 & 0 & 0 & 0 \\ 1 & 1 & 0 & 0 & 1 & 1 & 0 & 0 \\ 1 & 0 & 1 & 0 & 1 & 0 & 1 & 0 \\ 0 & 1 & 1 & 0 & 1 & 0 & 0 & 1 \end{pmatrix}$$

for $\mathcal{RM}(1,3)$. No set of one, two or three columns of H is linearly dependent, but $\mathbf{c}_1 + \mathbf{c}_2 + \mathbf{c}_7 + \mathbf{c}_8 = \mathbf{0}$, so $\mathcal{RM}(1,3)$ has minimum distance $d = 4$.

Bibliography

[An74] I. Anderson, *First Course in Combinatorial Mathematics*, Oxford University Press, Oxford, 1974.

[As65] R. Ash, *Information Theory*, Wiley, New York, 1965.

[Ba63] G. Bandyopadhyay, A simple proof of the decipherability criterion of Sardinas and Patterson, *Information and Control* 6 (1963), 331–336.

[Be68] E. R. Berlekamp, *Algebraic Coding Theory*, McGraw-Hill, New York, 1968.

[Be74] E. R. Berlekamp (ed.), *Key Papers in the Development of Coding Theory*, IEEE Press, New York, 1974.

[BP85] J. Berstel and D. Perrin, *Theory of Codes*, Academic Press, Orlando, 1985.

[Bi65] P. Billingsley, *Ergodic Theory and Information*, Wiley, New York, 1965.

[BM75] I. F. Blake and R. C. Mullin, *The Mathematical Theory of Coding*, Academic Press, New York, 1975.

[BM76] I. F. Blake and R. C. Mullin, *Introduction to Algebraic and Combinatorial Coding Theory*, Academic Press, New York, 1976.

[BR98] T. S. Blyth and E. F. Robertson, *Basic Linear Algebra*, Springer Undergraduate Mathematics Series, Springer, London, 1998.

[Br56] L. Brillouin, *Science and Information Theory*, Academic Press, New York, 1956.

[CL91] P. J. Cameron and J. H. van Lint, *Designs, Graphs, Codes and their Links*, LMS Student Texts 22, Cambridge University Press, Cambridge, 1991.

[Ch85] W. G. Chambers, *Basics of Communications and Coding*, Oxford University Press, Oxford, 1985.

[CS92] J. H. Conway and N. J. A. Sloane, *Sphere Packings, Lattices and Groups* (2nd ed.), Springer-Verlag, New York, 1992.

[De74] N. Deo, *Graph Theory with Applications to Engineering and Computer Science*, Prentice-Hall, Englewood Cliffs, 1974.

[Ev63] S. Even, Tests for unique decipherability, *IEEE Trans. Information Theory* IT-9 (1963), 109–112.

[Ev79] S. Even, *Graph Algorithms*, Pitman, London, 1979.

[Fe50] W. Feller, *Introduction to Probability Theory and its Applications*, I, Wiley, New York, 1950.

[Fi83] E. Fisher, *Intermediate Real Analysis*, Springer-Verlag, New York, 1983.

[Gi52] E. N. Gilbert, A comparison of signalling alphabets, *Bell System Tech. J.* 31 (1952), 504–522.

[GM59] E. N. Gilbert and E. F. Moore, Variable-length binary encodings, *Bell System Tech. J.* 38 (1959), 933–967.

[Go49] M. J. E. Golay, Notes on digital coding, *Proc. IEEE* 37 (1949), 657.

[Go80] S. W. Golomb, Sources which maximise the choice of a Huffman coding tree, *Information and Control* 45 (1980), 263–272.

[Go88] V. D. Goppa, *Geometry and Codes*, Kluwer, Dordrecht, 1988.

[Ha67] M. Hall, Jr, *Combinatorial Theory*, Blaisdell, Waltham Mass., 1967.

[Ha48] R. W. Hamming, Single error-correcting codes — Case 20878, *Memorandum* 48-110-52, Bell Telephone Laboratories, 1948.

[Ha50] R. W. Hamming, Error detecting and error correcting codes, *Bell System Tech. J.* 29 (1950), 147–160.

[Hi86] R. Hill, *A First Course in Coding Theory*, Oxford University Press, Oxford, 1986.

[Hu52] D. A. Huffman, A method for the construction of minimum redundancy codes, *Proc. IRE* 40 (1952), 1098–1101.

[Jo79] D. S. Jones, *Elementary Information Theory*, Oxford University Press, Oxford, 1979.

[Ka61] J. Karush, A simple proof of an inequality of McMillan, *IRE Trans. Information Theory* IT-7 (1961), 118.

[Ke56] J. L. Kelley, Jr, A new interpretation of information rate, *Bell System Tech. J.* 35 (1956), 917–926.

[KR83] K. H. Kim and F. W. Roush, *Applied Abstract Algebra*, Ellis Horwood, Chichester, 1983.

[Kn73] D. E. Knuth, *The Art of Computer Programming, vol. I: Fundamental Algorithms*, Addison-Wesley, Reading Mass., 1973.

[Kr49] L. G. Kraft, A device for quantizing, grouping, and coding amplitude modulated pulses, M. S. thesis, Electrical Engineering Department, MIT, 1949.

[La83] S. Lang, *Undergraduate Analysis*, Springer-Verlag, New York, 1983.

[Le64] V. I. Levenshtein, Some properties of coding and self adjusting automata for decoding messages, *Problemy Kiberneticki* 11 (1964), 63–121. (Russian)

[Li82] J. H. van Lint, *Introduction to Coding Theory*, Springer-Verlag, New York, 1982.

[LG88] J. H. van Lint and G. H. van der Geer, *Introduction to Coding Theory and Algebraic Geometry*, DMV Seminar 12, Birkhäuser, Basel, 1988.

[McE77] R. J. McEliece, *The Theory of Information and Coding*, Encyclopedia of Mathematics and its Applications 3, Addison-Wesley, Reading Mass., 1977.

[McM56] B. McMillan, Two inequalities implied by unique decipherability, *IRE Trans. Information Theory* IT-2 (1956), 115–116.

[MS77] F. J. MacWilliams and N. J. A. Sloane, *The Theory of Error-Correcting Codes*, North-Holland, Amsterdam, 1977.

[Mu53] S. Muroga, On the capacity of a discrete channel, *J. Phys. Soc. Japan* 8 (1953), 484–494.

[Pl82] V. Pless, *Introduction to the Theory of Error-Correcting Codes*, Wiley, New York, 1982.

[PH98] V. S. Pless and W. Huffman (eds), *Handbook of Coding Theory* (2 vols), Elsevier, Amsterdam, 1998.

[Pr92] O. Pretzel, *Error-Correcting Codes and Finite Fields*, Oxford University Press, Oxford, 1992.

[Re61] F. M. Reza, *An Introduction to Information Theory*, McGraw-Hill, New York, 1961.

[Ri67] J. A. Riley, The Sardinas-Patterson and Levenshtein theorems, *Information and Control* 10 (1967), 120–136.

[SP53] A. A. Sardinas and C. W. Patterson, A necessary and sufficient condition for the unique decomposition of coded messages, *IRE. Internat. Conv. Rec.* 8 (1953), 104–108.

[Sc64] E. S. Schwartz, An optimum encoding with minimum longest code and total number of digits, *Information and Control* 7 (1964), 37–44.

[Sh48] C. E. Shannon, A mathematical theory of communication, *Bell System Tech. J.* 27 (1948), 379–423, 623–656.

[SW63] C. E. Shannon and W. Weaver, *The Mathematical Theory of Communication*, University of Illinois Press, Urbana, 1963.

[Se67] D. A. R. Seeley, A short note on Bandyopadhyay's proof of the decipherability criterion of Sardinas and Patterson, *Information and Control* 10 (1967), 104–106.

[Si64] R. C. Singleton, Maximum distance q-nary codes, *IEEE Trans. Information Theory* IT-10 (1964), 116–118.

[Sl74] D. Slepian (ed.), *Key Papers in the Development of Information Theory*, IEEE Press, New York, 1974.

[St93] H. Stichtenoth, *Algebraic Function Fields and Codes*, Springer-Verlag, Berlin, 1993.

[ST81] M. N. S. Swamy and K. Thulasiraman, *Graphs, Networks and Algorithms*, Wiley, New York, 1981.

[Th83] T. M. Thompson, *From Error-Correcting Codes through Sphere Packings to Simple Groups*, Carus Mathematical Monographs 21, Math. Assoc. of America, 1983.

[Va57] R. R. Varshamov, Estimate of the number of signals in error correcting codes, *Dokl. Akad. Nauk SSSR* 117 (1957), 739–741. (Russian)

[We88] D. Welsh, *Codes and Cryptography*, Oxford University Press, Oxford, 1988.

[Zi59] S. Zimmerman, An optimal search procedure, *Amer. Math. Monthly* 66 (1959), 690–693.

Index of Symbols and Abbreviations

The symbol □ is used in the text to mark the end of a proof. The following symbols, in regular mathematical use, are used without further comment:

C	the set of complex numbers
R	the set of real numbers
Q	the set of rational numbers
Z	the set of integers
N	the set of natural numbers $\{1, 2, \ldots\}$
$\mathbf{Z}_n$	the set of integers mod (n)
$[a, b]$	the set of all real numbers x satisfying $a \leq x \leq b$
$(a, b]$	the set of all real numbers x satisfying $a < x \leq b$
(a, b)	the set of all real numbers x satisfying $a < x < b$
S^n	the set of ordered n-tuples from a set S
$A \setminus B$	the set of all elements lying in A but not in B
$\emptyset$	the empty set
$|S|$	the size of the set S
$n!$	factorial n $(= 1.2.3 \ldots n)$
$\binom{n}{r}$	the binomial coefficient $(= n!/r!\,(n-r)!)$
$\approx$	is approximately equal to
$\equiv$	is congruent to
$\log a$	the logarithm of a (to some unspecified base)
$\log_r a$	the logarithm of a to the base r
$\lg a$	$\log_2 a$
$\ln a$	$\log_e a$
∞	infinity
$\rightarrow$	tends towards, or approaches
$\mapsto$	is mapped to

$f'(x)$	the derivative of the function $f(x)$	
$\wedge, \vee$	the logical connectives "and" and "or"	
$\cap, \cup$	intersection and union	
$\sum$	sum	
$\prod$	product	
(a_{ij})	the matrix with entry a_{ij} in the (i, j) position	
$\det(A)$	the determinant of a matrix A	
$\text{tr}(A)$	the trace of a matrix A	
A^{T}	the transpose of a matrix A	
I_n	the $n \times n$ identity matrix	
$\mathbf{a.b}$	the scalar or dot product of vectors $\mathbf{a}$ and $\mathbf{b}$	
$\Pr(X_n = s_i)$	the probability that a variable X_n takes the value s_i, also written as $\Pr(s_i)$	
$\Pr(b\|a)$	the probability of b given a	
min	minimum	
max	maximum	
$\lfloor x \rfloor$	the integer part of x, the greatest integer $i \leq x$	
$\lceil x \rceil$	the "ceiling function", the least integer $i \geq x$	

The following symbols are defined in the text on the page indicated, and are then used without comment.

Index

AAO-6779